T0073068

Dialogues in Climate and Environmental Research, Policy and Planning:
A Special Focus on Zimbabwe

Edited by
Innocent Chirisa

Langaa Research & Publishing CIG
Mankon, Bamenda

Publisher
Langaa RPCIG
Langaa Research & Publishing Common Initiative Group
P.O. Box 902 Mankon
Bamenda
North West Region
Cameroon
Langaagrp@gmail.com
www.langaa-rpcig.net

Distributed in and outside N. America by African Books Collective
orders@africanbookscollective.com
www.africanbookscollective.com

ISBN-10: 9956-551-16-3

ISBN-13: 978-9956-551-16-3

© Innocent Chirisa 2020

All rights reserved.
No part of this book may be reproduced or transmitted in any form or by any
means, mechanical or electronic, including photocopying and recording, or be
stored in any information storage or retrieval system, without written permission
from the publisher

Notes on Contributors

Angeline Maturure is a Geography Tutor and Geography Teacher at Dzivarasekwa High 1. She holds an MSc in Development Studies with Women's University in Africa, a Bachelor of Arts with the University of Zimbabwe and a Diploma in Education. Her research interests are in Climate Change and Gender.

Aurthur Chivambe is a Graduate of BSc Hons in Rural and Urban Planning, (University of Zimbabwe). His research interests include socio-economic development, project planning and management and urban and regional planning for sustainability.

Chipo Mutonhodza is a lecturer in the Department of Urban Planning and Development at Great Zimbabwe University. She holds a MSc Development Studies (Women's University in Africa) and a BSc Rural and Urban Planning from University of Zimbabwe. Currently she is studying towards an MSc Rural and Urban Planning at University of Zimbabwe. Her research interests are in spatial planning and disaster risk reduction, cultural heritage and urban planning as well as transport planning. She has six years' experience working as a Town Planner in the Ministry of Local Government, Department of Physical Planning.

Conillious Gwatirisa is DPhil Student with the University of Zimbabwe's Centre for Applied Social Sciences specialising in Public Policies (Public Sanitary facilities).He is currently engaged as a Part-Time Lecturer with Women's University in Africa lecturing the Environment and Sustainable Development Course. He holds the following degrees with the University of Zimbabwe; an MSc in Environmental Policy and Planning, a BSc Hons (Geography),a B.Ed (Geography) and a Diploma in Education (Geography).His research interests are in Water and Sanitation, Climate Change, Gender and Public Policies.

Elmond Bandauko holds a Master of Public Administration (MPA) with specialisation in local government from the University of Western Ontario (Canada) where he studied as an African Leaders of Tomorrow Scholar. He did his BSc. (Hons) in Rural and Urban Planning from the University of Zimbabwe. His interests include participatory policy-making, policy innovation and policy diffusion, public management, programme and policy evaluation, collaborative governance and the politics of urban development in cities of the global south. He has published 18 book chapters, 18 journal articles and is a co-editor to a book on peri-urban developments and processes in Africa with Special Reference to Zimbabwe which was published by Springer International Publishing in 2016. He looks forward to pursuing his PhD in policy studies or in urban planning, focusing on city-regions, economic development and social planning.

Emma Maphosa is urban and regional planner who graduated with a BSc Honours Degree in Regional and Urban Planning with the University of Zimbabwe in 2018. In 2017, Emma Maphosa was employed with Gutu Rural District Council in Masvingo Zimbabwe. Her research interests are in real estate property management, urban management and transport planning.

Gladys Mandisvika is a holder of a BSc Honours in Rural and Urban Planning of the Department of Rural and Urban Planning, University of Zimbabwe. She once worked as a Graduate Teaching Assistant in the same department. Her research interests are in urban development, transport planning and management and sustainable cities.

Halleluah Mangombe-Chirisa is a holder of A BScEd dual honour's in Geography and Mathematics from the Bindura University of Science Education (BUSE) and an MSc degree in Population Studies from the University of Zimbabwe (UZ). Between 2005 and 2007, she taught at Nyava and Rutope secondary schools in Shamva district and Glen View 3 High School in Harare. Currently, she is a social scientist at Population Services International (PSI) - Zimbabwe in Harare. She has also worked as an independent consultant for the Zimbabwe Statistical Agency (ZIMSTAT), involved in national census exercises. Her research interests are population issues (especially reproductive health), resources mobilisation for communities and sustainability and resilience in urban and regional areas.

Innocent Chirisa is a full professor in the Department of Rural and Urban Planning. Since 2007, he has been researching on urbanisation, environment and housing among other related subjects. He completed his doctoral thesis on the stewardship of peri-urban areas and housing in Zimbabwe at the University of Zimbabwe in

2013. He has authored many articles in international peer-reviewed journals as well as book chapters. The articles and chapters cover an array of issues in urban planning and design, place stewardship, peri-urban settlements, community-based interventions and initiatives, social protection, urban policy, governance and management, women and gender debates, among other things. He has a special interest in the following research aspects: housing and shelter provision, planning ethics, planning education, social theory and planning, political theory and planning, disaster and risk management, environment stewardship and urban policy and governance. Currently, Professor Innocent Chirisa has been the Chairman of the Department of Rural and Urban Planning. Currently he is an Acting Dean of the Faculty of Social Studies at the University of Zimbabwe.

Liaison Mukarwi holds a BSc Honours Degree in Rural and Urban Planning from the University of Zimbabwe. He is currently practicing as a Town Planner at Human Settlements Experts (Pvt) Ltd in Zimbabwe. Liaison is a freelance researcher whose researches focus on Urban Management and Governance, Environmental Design, Urban Planning Practice, Housing and Community Issues and Transport and Sustainability issues. He had published more than several journal articles and book chapters. Liaison participated in various academic researches as a research assistant.

Nelson Chanza is a Senior Lecturer in the Department of Geography, Faculty of Science and Engineering, Bindura University of Science Education. He is also engaged as a Research Associate with the Department of Geosciences, Nelson Mandela University in South Africa. He holds a PhD in Environmental Geography from the Nelson Mandela University, South Africa. He currently serves as a Board of Director of the Environment Management Board under the Ministry of Environment, Tourism and Hospitality Industries in Zimbabwe. His research contributions are largely in the area of citizen science in environmental issues, in particular, indigenous knowledge applications in climate change impact assessment, adaptation and mitigation strategies that are used by local communities. He also actively participates in national and international training and capacity building on mainstreaming climate change in development programmes and disaster management programming.

Patience Mazanhi is a student studying towards an honour's in BSc Rural and Urban Planning at the University of Zimbabwe. Her research interests are in project planning, policy and project evaluation, land rights and social research.

Queen Linda Chinozvina is a holder of a BSc Honours Degree in Rural and Urban Planning with the University of Zimbabwe and currently pursuing a Master's Degree in Rural and Urban Planning with the same institution. She is also a Graduate Teaching Assistant at the Department of Rural and Urban Planning tutoring different courses. She has also worked as a research assistant for different organisations.

Spiwe Nyasha Mupukuta holds a BSc Honours in Rural and Urban Planning from the University of Zimbabwe. Her research interests are sustainable transport and urban environment dynamics.

Thomas Karakadzai is a Masters student at the Department of Rural and Urban Planning, University of Zimbabwe. He holds a BSc honours in Rural and Urban Planning from the University of Zimbabwe. His research interests are governance, sustainable human settlements and development issues.

Tinashe Bobo is a Town Planner working with the City of Harare in the City Planning and Development Division. Currently, He is studying towards the MSc in Urban Planning in the Department of Rural and Urban Planning at the University of Zimbabwe. He holds a BSc Honours Degree in Rural and Urban Planning from the University of Zimbabwe. His research interests are urban informalities, land-use planning, housing development, urban policy dynamics and management.

Verna Nel qualified as a town and regional planner at Wits University, and later obtained an MSc and a PhD through UNISA. She began her career at the Johannesburg Municipality and later worked in a private firm and national government before joining the Centurion Town Council. In 1998, she was appointed chief Town Planner at the municipality. She managed the City Planning function of the City of Tshwane from July 2001 to June 2008. In 2009, Verna moved to the Urban and Regional Planning Department of the University of the Free State as a Professor. She is an active researcher in addition to supervising Masters and doctoral students and teaching responsibilities. Her current research interests are urban resilience and spatial governance. She is also part of the South African Planning Education Research Project. Verna also serves on the South African Council for Planners and is the chairperson of the Education and Training Committee.

Wendy Tsoriyo is a Lecturer in the Department of Urban Planning and Development at Great Zimbabwe University. She has lecturing experience in Zimbabwe as well as well as training experience in community capacity building and development in Zimbabwean marginalized communities. Wendy is currently studying for a PhD in Urban and Regional Planning. Her research interests are in areas of spatial transformation of urban areas, participatory planning and urban design and management. Wendy enjoys volunteering her services for the betterment of her community and spending time with her children who are also her source of inspiration.

Zebediah Muneta is a student studying honours BSc in Real Estate Management at the University of Zimbabwe. His research interests are in land management, property valuation and management of real estate investments.

Table of Contents

Dedication

This book is dedicated to people who have access, use and control the environment and who acknowledge climate change as a challenge in our times.

Acknowledgements

The editor is indebted to extend gratitude to all the authors of chapters who worked tirelessly to have their chapters polished to the product presented in this volume. I would also want to thank the anonymous reviewers and the technical editors. Last, I want to thank everyone who contributed with the data that the chapter contributors used to convey their messages.

Chapter 1

Setting the Tone - Climate Research and Environmental Policy and Planning Debates

Innocent Chirisa

Introduction

Climate change is a long-term shift in the statistics of the weather (Boden, Marland and Andres, 2011). For example, it could show up as a change in climate normal (expected average values for temperature and precipitation) for a given place and time of year, from one decade to the next (Petit *et al.* 1999). Certain naturally occurring gases, such as carbon dioxide (CO_2) and water vapour (H_2O), trap heat in the atmosphere causing a greenhouse effect. Burning of fossil fuels, like oil, coal and natural gas is adding CO_2 to the atmosphere. There are numerous potential effects of climate change. Extensive research is being done around the world a good deal within National Climatic Data Centre (NOAA) to determine the extent to which climate change is occurring, how much of it is being caused by anthropogenic or manmade forces and its potential impacts. It is now more certain than ever, based on many lines of evidence, that humans are changing Earth's climate. The atmosphere and oceans have warmed, accompanied by sea-level rise, a strong decline in Arctic sea ice and other climate-related changes.

This chapter seeks to discuss the vulnerability of cities to many types of shocks and stresses, including natural hazards. These include storms and sea level rise, but also man-made ones like economic transformation and rapid urbanisation. The chapter enhances the discourse that realise lack of consensus in some of these areas, among scientists and the often-conflicting points-of-view and studies. However, with further research, no doubt many questions regarding impacts will be resolved in the future. Potential impacts most studied by researchers include the effects on sea level, drought, local weather and hurricanes. This chapter discusses that these shocks and stresses have the potential to bring cities to a halt and reverse years of socio-economic development gains. Cities that

1

are to grow and thrive in the future must take steps to address these shocks and stresses (Herbert, 2013). Simply put, a resilient city is one that can adapt to these types of changing conditions and withstand shocks while still providing essential services to its residents. The major dispute is on how the issue of both climate and environmental research come into play when different stakeholders pursue the considering that planning involves three operational lag time; short term, medium term and long term.

The chapter recommends that information is the foundation of sustainable development and is fundamental to successful planning and decision-making. Satellite and other remote sensing technologies can improve capabilities of capturing environmental status and trends. In addition, satellite imagery, geographic information systems (GIS) and aerial photography have greatly expanded opportunities for data integration and analysis, modelling and map production. There is need for enhanced environmental research, climate research, training and dissemination of environmental management tools in Africa.

Contextual and Theoretical Underpinnings

India is one of the world's most vulnerable countries to climate change (Cruz *et al.* 2007). About half of India's population is dependent on agriculture or other climate sensitive sectors (Bureau of Labour Statistics, India, 2010). Also, about 76 percent of the Indian population lives on less than $2 a day (Boden 2011). UNDP (2010) found that poverty levels in eight Indian states are as acute as those in the 26 poorest countries in Africa. These eight Indian states are home to about 421 million people in poverty, 11 million more people in poverty than in the 26 poorest African countries combined. India is vulnerable to sea level rise and extreme weather events and will increasingly face threats to human health, water availability and food security (Cruz *et al.* 2007). Additionally, about 12 percent (40 million hectares) of India is flood prone, while 16 percent (51 million hectares) is drought prone (Central Water Resources, 2011). Consequently, India is also vulnerable to potential climate change induced shifts in precipitation patterns.

Climate change and energy are now a focus of local, state and national attention around the world. India has long been a key player in international negotiations. It has begun implementing a diverse portfolio of policies nationally and within individual states to improve energy efficiency, develop clean sources of energy and prepare for the impacts of a changing climate (Cruz *et al.* 2007). An effective national strategy, however, must consider the climate change and energy-related beliefs, attitudes, policy preferences and behaviours of the Indian people. They play a vital role in the success or failure of this strategy through their decisions as citizens, consumers and communities. Building public acceptance, support and demand for new policies to both limit the severity of global warming and prepare for the impacts of a changing climate. It will require education and communication strategies based upon a clear understanding of what Indians already know, believe and support, as well as what they currently misunderstand, disbelieve, or oppose.

The European environment policy rests on the principles of precaution, prevention and rectifying pollution at source and on the 'polluter pays' principle. Multiannual environmental action programmes set the framework for future action in all areas of environment policy. They are embedded in horizontal strategies and taken into account in international environmental negotiations. Last but not least, implementation is crucial. The precautionary principle is a risk management tool that may be involved in case of scientific uncertainty about a suspected risk to human health or to the environment emanating from a certain action or policy. For instance, certain individual projects (private or public) with prospective significant effects on the environment, such as motorway or an airport construction, are subject to an environmental impact assessment (EIA) (UNDP, 2010). Equally, a range of public plans and programmes (e.g. concerning land use, transport, energy, waste or agriculture) are subject to a similar process called a strategic environmental assessment (SEA). Here, environmental considerations are already integrated at the planning phase and consequences are considered before a project is approved or authorised so as to ensure a high level of environmental protection. In both cases, consultation with the public is a central aspect.

Now for the first time in history, more than half the world's population lives in urban areas (WHO, 2010). Over 90 percent of urbanisation is taking place in developing countries, with the African Region experiencing the highest rate of urbanisation at 3.5 % annually (UNEP). With energy use in urban electricity, transport and industry contributing up to 77 percent of Global Greenhouse Gas (GHG) emissions, cities constitute a major source of GHG emissions (Satterthwaite, 2008). But not only are cities significant contributors to climate change, they are also most likely to disproportionately suffer its consequences on account of their proximity to coastlines or water bodies and are, therefore, susceptible to flooding and drought (Nicholls *et al.* 2007). Climate-change induced floods are already having very large impacts on urban centres in many African nations. For instance, heavy floods affected Maputo, Mozambique in 2000, floods and mudslides brought heavy damage to urban East Africa in 2002. There are also serious floods in Port Harcourt, Nigeria and in Addis Ababa, Ethiopia in 2006 displaced tens of thousands of people.

Scientists across the globe have stated that warming of the climate system is unequivocal and that it is largely spurred by human activities releasing greenhouse gases (GHG) into the atmosphere (National Environment Policy, 2013:33). For Kenya, climate change poses many serious and potentially damaging effects on human and the environment in the coming decades. Increased vulnerability in climate and projected incremental changes associated with air and sea temperature, precipitation and sea level and changes in the frequency and severity of extreme events will have profound social, economic and ecological implications. The effects of climate change have the potential to disrupt Kenya's strides in agricultural production, forests, water supply, health systems and overall human development.

Climate change, projected to exacerbate existing desertification and water stress, constitutes a major threat to Northern Africa's urban populations (Herbert, 2013). The negative effects on agriculture will increase the need for food imports, with adverse effects on regional balances of payments. This might even restart the rural-urban exodus that seemed to have abated some years ago. Northern Africa's cities are largely dependent on the recharge of huge aquifers under the Sahara Desert, but these groundwater sources are being depleted. For instance, along the Nile Valley the effects of climate change are expected to be particularly severe and, according to some projections, potentially catastrophic. Urban water infrastructures must be rehabilitated and maintained to eliminate unnecessary wastage and more investment in wastewater treatment and reuse is essential.

Despite being the foundation on which sustainable development is anchored, there are many environmental degradation issues and challenges facing the Kenyan country

4

(National Environment Policy, 2013). The environment has been an essential feature of Kenya's development trajectory. Yet, the country lacks a comprehensive environment policy and most of the environmental imperatives are captured in various development plans. (National Environment Policy, 2013) assets that notable drivers of environmental degradation are high rates of population growth, inappropriate technology, unsustainable consumption and production patterns, increased incidences of poverty and climate change. Further, urban environmental degradation, through lack of appropriate waste management and sanitation systems, industry and transport related pollution, adversely impact on air, water, soil quality and human health and well-being. Another major set of challenges arises from emerging global environmental concerns such as stratospheric ozone depletion and biodiversity loss. These have led to changes in the relationship between people and ecosystems. If this trend is left unchecked it will lead to further serious environmental degradation that may perpetuate deprivation and poverty.

It is important to note, the old Constitution of Zimbabwe (amended in 2000) did not have a specific clause that provided for the protection of the environment (Government of the Republic of Zimbabwe, 2000). It is against this background that in 2002, the government of Zimbabwe enacted the EMA (Chapter 20:27) and a draft National Environmental Policy in 2003 to provide legal specifications on how the environment would be protected. The EMA was a consolidated environmental legislative measure which was meant to be the overall environmental legislation in Zimbabwe. After the promulgation of the EMA in 2002, some acts that had to do with environmental management had to be repealed and incorporated into the EMA in order to ensure consistency with the social, economic and political demands of the country (Mukwindidza, 2008: 34). These were: The Natural Resources Act, 9 of 1996; the Atmospheric Pollution Prevention Act, 31 of 1996; the Hazards Substances and Articles Act, 76 of 1996; and the Noxious Weeds Act, 16 of 1993 (Mohammed-Katerere and Chenje, 2002: 54).

Following the enactment of the EMA in 2002, section 9 of this Act gave the (then) Minister of Environment and Tourism the power to establish an Environmental Management Agency whose duty was to formulate quality standards on air, water, soil noise, vibration, radiation and waste management. This Agency was formerly known as the Department of Natural Resources (Zimbabwe, 2014). The Agency is controlled and managed by the Environmental Management Board which is composed of experts from the areas of Environmental Planning and Management, Environmental Economics, Ecology, Pollution, Waste Management, Soil science, Hazardous substances as well as water and sanitation (Mukwindidza, 2008). The chapter outlines and explains the role of this Agency as follows:

- To develop guidelines for national plans, environmental management pans (EMPs) and local environmental action plans (LEAPS);
- To regulate, monitor, review and approve environmental impact assessments;
- To regulate and monitor the management and utilization of ecologically fragile ecosystems;
- To develop and implement incentives for the protection of the environment; undertaking
any works deemed necessary or desirable for the protection or management of the environment where it appears to be in the best interest of the public or where in the opinion of the Agency, the relevant authority has failed to do so;
- To serve written orders on any persons requiring them to undertake or adopt such measures as specified in the orders to protect the environment:
- To carry out periodic environmental audits of any projects including projects whose implementation started before the fixed date for the purpose of ensuring that their implementation complies with the requirements of the Act.

According to Chinamora (1995), Zimbabwe declared the EIA policy in 1994 and it now constitutes an essential tool for integrating environmental and economic considerations in the planning process (Chinamora, 1995: 153). During the time when Chinamora was writing his essay, the EIA was not yet law, but it became law soon after the enactment of the EMA of 2002 (Chapter 20:27) under statutory instrument 7 of 2007 (Environmental Impact Assessment and Ecosystems Protection) Regulations. This has compelled prescribed projects, listed under the first schedule of the

EMA Act (Chapter 20:27), to undergo an EIA process prior to implementation (Herald, 2012).

Broad public participation in decision-making processes is one of the fundamental preconditions for sustainable development. This presupposes access to timely and accurate information on the environment. Sound environmental management has to be based on openness and participation at all levels (Elala, 2011). Therefore, it is imperative that environmental education and public awareness is promoted at all levels. Extant are the Globe Scan Incorporated, the Yale Project on Climate Change Communication, in the Yale School of Forestry and Environmental Studies at Yale University. These conduct scientific research is on public environmental knowledge, risk perceptions, decision-making and behaviour and empowers educators and communicators with knowledge, training and tools. The major ideas to advance public understanding and engagement with climate change science and solutions. Henceforth, it is through environmental monitoring which is important for determining environmental status and trends and for updating environmental action plans and enhancing enforcement and compliance. It analyses many circumstances in which human activities carry a risk of harmful effects on the natural environment.

A resilient city can keep moving toward its long-term goals despite the challenges it meets along the way. Climate change will render most of Northern Africa's cities more vulnerable to disasters associated with extreme weather patterns, especially flooding, while desertification presents a threat to Sudan's rural economy and food production. Some governments have already responded by framing plans taking these threats into account. Careful monitoring and regional cooperation will help anticipate threats as they emerge and allow for exchange of ideas and information. The Nile Delta requires special attention as it is especially vulnerable to inundation and saltwater intrusion. On the other hand, Northern Africa's climate offers immense opportunities for the generation of renewable solar and wind energy and exploitation of these has commenced.

Most of our current knowledge of global change comes from General Circulation Models (GCMs). At present, GCMs have the ability to provide us with a mean annual temperature for the planet that is reliable. Regional and local temperature and precipitation information from GCMs is, at present, unreliable. Much of the global change research effort is focused on improving these models.

As part of the resilience agenda, Ethiopia plays a prominent role in the struggle to mitigate and adapt to climate change, as shown by the active engagement of the government and its collaboration with multiple actors to reduce the country's vulnerabilities. Climate change impacts in Ethiopia, such as the increase in average temperature and changes in rainfall distribution, exacerbates current vulnerabilities that are highly interlinked with other shocks and stresses such as rapid urbanisation (Elala, 2011). There are supporting the city in their pursuit of sustainable development it is a necessity as well as an opportunity for our country to embark on a resilient development path.

Environmental education, both formal and informal, is vital to changing people's attitude to appreciate environmental concerns (National Environment Policy, 2013: 44). Formal education is important to increase awareness, improve extension services, sensitive people on environmental issues and build institutional capacities whilst non-formal environmental education benefits people outside the formal education system (National Environment Policy, 2013: 44). Communication of environmental information to all stakeholders is still a challenge. Therefore, public awareness empowers the public to develop a strong sense of responsibility on environmental and climatic issues. The biggest factor in determining future global warming is projecting future emissions of carbon dioxide and other greenhouse gases. This in turn depend on how people will produce and use energy, what national and international policies might be implemented to control emissions and what new technologies might become available (National Research Council, 2010e). Scientists try to account for these uncertainties by developing different scenarios of how future emissions and hence climate forcing will evolve (National Research Council, 2011a). Each of these scenarios is based on estimates of how different socioeconomic, technological and policy factors will change over time, including population growth, economic activity, energy-conservation practices, energy technologies and land use.

Content Focus of the Book

Chapter 1, 'Setting the Tone' provides the background of the book and introduces the focus and thrust of the book.

Chapter 2, entitled 'The Climate-Change-Urbanisation Conundrum in Africa: A Regional Research Perspective' by

Halleluah Chirisa, Gladys Mandisvika, Elmond Bandauko and Nelson Chanza seeks to provoke a debate on climate science and policy, highlighting how the multi-method research approach provides a basis, not only for critical thinking, but also for appropriate action in addressing the complex issue of climate change effects on urbanisation. Such debate is long overdue for Africa, which is at the tipping point of both forces, with the highest rate of urbanisation; due largely to rural-to-urban migration and being the worst affected in terms of climate change induced impacts. Little has been done to quantify and factualise this reality. This is detrimental to informed local action and preparedness and to setting priorities in addressing the extremities of climate change and urbanisation. The present study seeks to fill the gap in practice by advocating strongly for the use of a combination of tools, strategies and methods to research issues in climate change and urbanisation in Africa. The study makes use of African case studies, content analysis of existing documents on the subject and statistics to demonstrate realities, gaps and possible trajectories to explain these phenomena.

Chapter 3, 'Climate Change and Women in Ward 16, Goromonzi' by Angeline T Maturure and Conillious Gwatirisa acknowledges Zimbabwe's recent experiences of severe effects of climate change in the form of erratic rainfall, heat waves and crop failure. Studies on Climate Change have found out that women bear the brunt of the acute impacts of climate change as they constitute the majority of the world's most economically disadvantaged. Women are disproportionately more involved in natural resource dependent activities than men. The main objective of this study was to establish the social-economic impacts of climate change on women. The study also sought to determine the coping strategies adopted by women in Ward 16 of Goromonzi. The study employed both quantitative and qualitative methods that included the administration of a household questionnaire, participatory observations and key informant interviews. The study established that women were indeed facing serious and selective effects of climate change as evidenced by critical water and food shortages. The study established that to mitigate the impacts of climate change some women resorted to the following strategies: marrying off their teenage girls, selling livestock and diversifying into cross border trading. Others engaged in anti-social behaviour such as engaging into sex work. A number of recommendations were proposed, and these included the provision of food aid, the introduction of cash transfers and income generating projects. Households were also being urged to improve their capabilities by diversifying into short-season crops as well as by engaging into conservation farming. Lastly, the government and development partners were being urged to construct small dams for irrigation as well as to rehabilitate the existing water infrastructure.

Chapter 4, 'Urban Environmental Planning and Disaster Preparedness by Innocent Chirisa, Sharon Marimira and Artwell Nyirenda: A Focus on Beitbridge Town' observe that the planning regulation and development control mechanisms and tools in Zimbabwe for most cities have so far made little effort to create disaster resilient cities. In an environment where climate change, urbanisation and human activities have accelerated the frequency and magnitude of natural disasters such as floods, cyclones and droughts, it is necessary to be proactive to enable our cities to become functional again after experiencing shocks and stresses. This chapter aims to explore the gaps and constraints in environmental planning and disaster preparedness in African urban areas, using Beitbridge as a case study. It uses secondary data and key informant interviews, which were conducted to gather information about floods in the study area. The study shows that the early warning systems and response mechanisms of local authorities to floods in the area are weak and that there is little or no proactive planning. The research recommends improved communication so that all potential victims can be reached, an effective coordinated flow of information between institutions and the constant monitoring and evaluation of the disaster preparedness plan, especially vulnerability assessment, to increase the level of responsiveness of the local authorities to disasters.

Chapter 5, 'Inside the Climate Change Policy-Making in Zimbabwe' by Kenneth Odero is premised on the argument that the presence (or absence) of national climate change policy frameworks in a given country is often taken as a testament of its capacity to formulate a climate change policies and response strategies. Using this logic, Zimbabwe's slow pace in formulating a national climate change policy and response strategy has been construed as symptomatic of a 'lack of capacity'. Such a binary view of climate policy-making is flawed for the following reasons. First, it fails to appreciate national policy-making interests and priorities. Second, it is ahistorical and totally neglects embedded policy-making capabilities that come with experience. Given Zimbabwe's deep policy-making know how and experience, explanation for its relatively slow pace in national climate change policy-making must lie elsewhere. This essay points to Zimbabwe's constricted policy space, disengagement with the international community and the contradictions around the climate change agenda as the real reasons for the 'slow' pace in formulation of a national climate change policy and response strategy.

Chapter 6, entitled 'Urban Climate Resilience by Chipo Mutonhodza, Patience Mazanhi and Aurthur Chivambe: A Review for Zimbabwe' acknowledge that Africa is urbanising fast and the challenges emanating from climate change have debilitating effects of the region and its people. This chapter is informed by a study that set to examine the extreme climatic conditions that Southern Africa and perhaps the rest of Africa which have seen water resources oscillating between copiousness and immense insufficiency. In recent times the El-Nino and La-Nina phenomena have challenged these urban spaces and hence accentuate the water stress and challenges that come by intense rainfall including flooding. In such a context, it becomes imperative that authorities running cities look into equilibrating factors, tools and techniques to ensure their inhabitants safety, comfort and reduced costs of habitation in that cities and that the situation does not culminate into an unnecessary urban penalty. The chapter uses literature and document review especially from Asia and Pacific whose experiences are rich in providing case studies and policy options those countries like Zimbabwe can learn from. As such, the chapter is informed by a desktop study which then places the urban Zimbabwe situation and trying to sift into the constraints and challenges that one has overcomes and proffer workable solutions including the climate proofing, water harvesting, green infrastructure development. One initiative which is often used by cities in Zimbabwe, of late, have been, water rationing or letting individual households drill own boreholes or resort to deep or shallow wells as a source of water. The sustainability of these techniques is questionable. In light of this uncertainty characterising these methods, the study draws lessons from Asia and Pacific.

Chapter 7, 'Climate Change Governance the Implications it has for Physical Planning in Zimbabwe' by Liaison Mukarwi and Wendy Tsoriyo seeks to bring an understanding of the terrain of climate change governance in Zimbabwe and implications that it has on physical planning as a public policy domain in which initiatives, innovations and instruments by state and non-state actors interact. Specific questions sought for answering are: What gaps exist in the instruments and innovations for implementation of climate change adaptation strategies? To what extent is physical planning a matter in the climate change adaptation strategies? The chapter uses the case method and narratology, which are methods of critical text analysis of existing documents by or about the important stakeholders, in situating the ideas, initiatives and programmes by the key stakeholders. The chapter acknowledges that proper societal transformation is a production of committed coordination of which meaningful climate change adaptation and mitigation is the foundation. Overall, in Africa, coordination is not a given and most of the initiatives in climate change adaptation, having a physical planning import, are often overlooked.

Chapter 8, 'Urban Planning Tools for Climate Risk Management in Zimbabwe' by Sharon Marimira, Chipo Mutonhodza and Thomas Karakadzai, seeks to analyse how urban planning tools in Zimbabwe manage or adapt to climate change so as to pave way to creating climate change resilient cities. Land use planning can be used to enable both local and national adaptation to climate change though official plans, zoning and development permits to minimise risks to communities from floods, wildfires, landslides and other natural hazards. This chapter is based on research done using literature and document review, which interrogated existing published and unpublished documents and databases such as newspapers and journals. Document review was particularly useful in providing estimates of relevant parameters such as current data about what is happening on the ground in terms of climate change adaptation in Zimbabwe.

Chapter 9, 'A Case for 'Retrofits' in the Urban Sector of Zimbabwe' by Emma Maphosa and Zebediah Muneta explains how the retrofits projects can be adopted in the contexts of Zimbabwe particularly as an option in addressing climate change challenges such as water shortages, energy crises, increases in global warming and greenhouse gas emissions, flood management flood management challenges and increase in waste management complications in all its sectors particularly the domestic sector residential sector. Retrofitting offers an opportunity to counter these challenges that have emanated as a result of rapid urbanisation and climate change among other factors by altering adding fitments

12

to existing building stock. By dividing retrofits into water, energy, carbon and flood retrofits and through examples of specific such projects, the chapter examines the success of these retrofit initiatives in global wide, regional and local initiatives. Challenges examined in implementing these projects locally are examined. From the examination of these domestic projects and a comparison with successful retrofits projects elsewhere in the world, the chapters propose recommendations for action to make these initiatives a successful in the Zimbabwean context.

Chapter 10, 'Glocalising International Environmental Law in Zimbabwe: Practices, Gaps and Direction' by Halleluah Chirisa, Aurthur Chivambe and Liaison Mukarwi is an attempt to evaluate what Zimbabwe, as a nation, has been doing in practice, identifying the gaps and proffer a nuanced and informed direction toward the protection of not only the global commons but also the local natural resources in a sustainable manner. This chapter provides an overview of the trends in the evolution of environmental law over the period, the varying rationale and institutional frameworks adopted in Zimbabwe, the evolving challenges, the nature of the institutional and regulatory frameworks for environmental law in the sub-region, global and the place and role of environmental law in promoting sustainable urban development in Zimbabwe in the context of the prevailing dominance of informal economic activities and increasingly uncontrollable urban sprawl. It finally assesses the effectiveness of the various planning approaches adopted, environmental policies, tools (Environmental Impact Analysis (EIA), Strategic Environmental Analysis (SEA) and organisations implemented by the government of Zimbabwe and also evaluate the relevance, impact and efficacy of international environmental law, in the context of the overall sustainable urban development management in Zimbabwe.

Chapter 11, 'The Food-Water-Health-Energy-Climate Change Nexus: Pivot for Resilience in the Cities of the Global South' by Innocent Chirisa, Verna Nel and Romeo Dipura explores the nexus between water, food, health, energy and climate change, as well as, the possibility of capitalising on synergies in perpetuating sustainability in countries of the Global South. It explores the complex relationship between health, food, water and energy systems as well as the effects of climate change on the nexus between these component systems. It focuses on reviewing the integrated system that supplies energy, water and food and its impact on health in the context of the developing world. The ability to ensure food security in developing countries depends on the understanding of risks and the vulnerability to climate change; particularly from the perspective of the nexus between food, water, health and energy. For methodology this study takes an exploratory research approach which is based entirely on secondary data sources. It examines the potential of the nexus approach in promoting sustainable development and transforming the vulnerable communities of the Global South into resilient societies. Integrated models for the monitoring and evaluation of the health-food-water-energy nexus are recommended in this chapter.

Conclusion

This chapter has demonstrated that individuals, policymakers and societies must be aware of and have at least a basic understanding of a threat to make informed decisions about how to respond. Education level was one of the strongest predictors of climate change awareness. Inadequate and disaggregated data is a challenge for integrated planning and management of the environment and undertaking total economic valuation. The chapter concludes that coordinated and participatory approach to environmental protection and management should be enhanced to ensure that the relevant government agencies, county governments, private sector, civil society and communities are involved in planning, implementation and decision-making processes. Scientific research technology and innovation are central to sound environmental management and climate change. High quality data generated from environmental research, climate research and monitoring can improve the country's information base for decision-making on environmental issues. Communication of environmental information to all stakeholders is still a challenge in Africa.

References

Boden, T. A., Marland, G., and Andres, R. J. (2011). Global, Regional and National Fossil-Fuel CO_2 Emissions. Carbon Dioxide Information Analysis Centre, Oak Ridge

Bureau of Labour Statistics., and Government of India. (2010). Report on Employment and Unemployment Survey (2009–10). Bureau of Labour Statistics, Govt. of India. Available online: http://labourbureau.nic.in/Final_Report_Emp_Unemp_2009_10.pdf [Accessed 12 December 2016].

Central Water Resources (CWC). (2011). India: Country paper on Water Security. Available online: http://www.indiaenvironmentportal.org.in/reports-documents/ india-country-paper-water-security [Accessed 10 December 2016].

Chinamora, W. (1995). Zimbabwe's Environment Impact Assessment Policy of 1994: Can it achieve sound environmental management? *Zambezia*, 22(2), 153-163.

Cruz R. V., Harasawa, H., Lal, M., Wu, S., Anokhin, Y., Punsalmaa, B., Honda, Y., Jafari, M., Li, C., and Huu Ninh, N. (2007). Asia. Climate Change 2007: Impacts, Adaptation and Vulnerability. Contribution of Working Group II to the Fourth Assessment Report of the Intergovernmental Panel on Climate Change. Available online: http://www.ipcc.ch/publications_and_data/ar4/wg2/en/ch10.html [Accessed 12 December 2016].

Elala, D. (2011). Vulnerability assessment of surface water supply systems due to climate change and other impacts in Addis Ababa, Ethiopia. Academic Thesis. Uppsala: Uppsala University

Government of the Republic of Zimbabwe. (2000). Constitution of the Republic of Zimbabwe. Harare: Zimbabwe

The Herald. (2012). "Zimbabwe: What Does the Law Say About Environmental Impact Assessment?" Harare: Zimpapers

Herbert, S. (2013). Assessing Seismic Risk in Ethiopia. Research Report. Birmingham: GSDRC, University of Birmingham.

UNDP. (2010). Human Development Report, United Nations Development Programme (UNDP). UNDP. Available online: http://hdr.undp.org/en/ [Accessed December 12, 2016].

UNDP. (2012). Human Development Report, United Nations Development Programme (UNDP). UNDP. Available online: http://hdr.undp.org/en/ [Accessed December 12, 2016].

Mohammed-Katerere, J. C., and Chenje, M. (2002). Environmental law and policy in Zimbabwe. Harare: Southern African Research and Documentation Centre, p. 54

Mukwindidza, E. (2008). The Implementation of Environmental Legislation in Mutasa District of Zimbabwe. Unpublished

Master of Public Administration Degree. Pretoria: University of South Africa, p. 34.

National Environment Policy. (2013). Ministry of Environment, Water and Natural Resources. Nairobi.

National Research Council. (2011a). Climate Stabilization Targets: Emissions, Concentrations and Impacts for Decades to Millennia

National Research Council. (2010e). Ocean Acidification: A National Strategy to Meet the Challenges of a Changing Ocean

Nicholls, R. J., Hanson, S., Herweijer, C., Patmore, N., Hallegatte, S., Jan Corfee-Morlot, J. C., and Muir-Wood, R. (2007). Ranking of the world's cities most exposed to coastal flooding today and in the future, OECD Environment Working Paper No. 1, 2007.

NOAA. National Climatic Data Centre site on Global Warming: Available online: http://www.ncdc.noaa.gov/oa/climate/globalwarming.html. [Accessed on 25 April 2017]

Petit, J. R., Jouzel, J., Raynaud, D., Barkov, N. I., Barnola, J. M., Basile, I., and Delmotte, M. (1999). Climate and atmospheric history of the past 420,000 years from the Vostok ice core, Antarctica. *Nature*, 399(6735), 429-436.

Satterthwaite, D. (2008). Cities' contribution to global warming: notes on the allocation of greenhouse gas emissions. *Environment and urbanization*, 20(2), 539-549.

WHO. (2010). UNEP, Africa Environment outlook, Past, Present and Future Perspectives Available online: http://www.unep.org/dewa/Africa/publications/aeo-1/203.htm. [Accessed on 25 April 2017]

Zimbabwe. (2014). *"Constitutional Requirement for Environmental Protection in Zimbabwe."* Handbook on Environmental Assessment Legislation in the SADC Region. Harare: Zimbabwe.

Chapter 2

The Climate-Change-Urbanisation Conundrum in Africa: A Regional Research Perspective

*Halleluah Chirisa, Gladys Mandisvika,
Elmond Bandauko & Nelson Chanza*

Introduction

Africa is severely confronted by the twin forces of climate change and urbanisation (UNEP, 1999). It is experiencing the highest rate of urbanisation, largely due to rural-to-urban migration and is the worst affected in terms of climate change induced impacts (Banda, 2013). Significantly, the manner in which climate change and urbanisation narratives have been conveyed in literature leaves much to be desired (Fiondella, 2011). Overall, the facts about and magnitude of the reality of climate change and urbanisation are absent from literature (Joubert, 2001; Bultitude, 2011; Fiondella, 2011). This is detrimental to informed local action and preparedness and to setting priorities in addressing the forces of climate change and urbanisation. The purpose of this chapter is to fill the gap in practice by advocating strongly for the use of a combination of tools, strategies and methods in researching issues about climate change and urbanisation in Africa. It seeks to provoke a critical scientific and policy debate, examining the extent to which the multi-method approach in research provides a basis for critical thinking and appropriate action in addressing the complex issue of urbanisation due to climate change effects.

This chapter is based on a desktop study, which makes use of African case studies, content analysis of existing documents on the subjects as well as statistical figures in demonstrating the realities, gaps and possible trajectories in explaining climate change and urbanisation researches. Document analysis is a social research method, which looks at reading materials such as public records, the media and visual documents (Bryman, 2001). The rationale for using document review was the nature of the analysis required in compiling this chapter which is about getting down to the historical and policy trends in the discussion of the nexus between climate change and urbanisation in Africa. The chapter begins by introducing discourse on the twin challenges of urbanisation and climate change in Africa. The second section tackles the contextual influences on this research on the issues for climate change and urbanisation in Africa. The third section discusses the research results and matches them with theoretical issues to come up with recommended policy options. The final section concludes the chapter.

Context and Background

It has become a common trend worldwide that more people now live in cities than in rural areas. However, African nations are experiencing urbanisation at more accelerated rates than other continents. The United Nations Population Fund (2007) estimates that most new urban population intensification will occur in smaller towns and cities, which have fewer resources to respond to the challenges of climate change. The UN (2014) projects that Africa's urban population will soar from 414 million to 1.2 billion people by 2050. Although there are many factors contributing to rural to urban migration, climate change has also been identified as a major push factor. According to IPCC (2014), migration owing to climate change is a complex issue largely because migrations in areas impacted by climate change are not unidirectional and permanent, but multi-directional and often temporary or episodic.

The reasons for migration are often multiple and complex and do not relate straightforwardly to climate variability and change. For instance, rural areas in Nigeria face problems of intensive drought, flooding and coastal erosion, which affect food production (Penning-Rowsell et al. 2013). People are then forced to move to urban areas in search of employment opportunities in tertiary and manufacturing services (UNEP, 1999). Interestingly, Banda (2013) observes that there is no universal understanding or agreement on the term to describe this kind of migration caused by climate change. Some refer to it as climate refugee-ism, while others call it climate-migration (IOM, 2008). Throughout much of Africa, climate-migration is driving urbanisation. Banda (2013) further elucidates that climate-migration is the forced dislocation of individuals or groups by sudden or gradual unfavourable changes in their environment. These changes lead to harsh conditions such as rising sea levels, which erode the land beneath coastal communities, desertification of farmland and flooding, which negatively affect living conditions.

The major threat to human survival and security is water scarcity, which is caused by rising temperatures. According to the IPCC (2013), Africa is experiencing temperature rises of roughly 0.7°C and it is predicted that temperatures will rise even further. UNESCO (2013) posits that almost 40 percent of Africans live in water-scarce environments and it is estimated that by 2030, 24 million people would have been displaced as water scarcity increases. Even if the realities of climate change induced urbanisation are evident, most governments see migration as a problem and something to discourage, while the migrants themselves regard movement as a form of adaptation to climate change. Proponents of urbanisation argue that the high level of urbanisation can boost economic growth. However, critics argue that colossal growth of urban populations can place a strain on the limited resources available in a city and can further aggravate existing stresses (Rondinelli *et al.* 1983). Climate change is expected to further increase the number of Africans living in slums. UNHABITAT (2010) has reported that in 2010, 61.7% of sub-Saharan Africa's urban population were slum dwellers and this is expected to increase with climate change. Banda (2013) elucidates that most African slums are crowded and usually situated on low lying ground, making the settlements more prone to flooding. Floods pose a very deadly human danger, as they are accompanied with air pollution, poor sanitation and increased vulnerability to malnutrition, disease and deaths.

Overall, the African region is experiencing the highest rate of urbanisation. This is largely due to rural-to-urban migration. It becomes the worst affected in terms of climate change induced impacts. In spite of this, the climate change and urbanisation narratives have been partially excluded (for example, UNHABITAT, 2010; Bultitude 2011; Fiondella 2011). With the little evidence on the impacts of climate change and urbanisation available, it is difficult for governments to make informed local action and preparedness and set priorities for addressing the impact of these two forces. It is therefore the aim of this chapter to debate the use of multi-method approach for providing sufficient answers that can inform local decision-making and action to address these two hostile forces. The next section describes the various methods that can be used in combination to deal with climate change and urbanisation.

Analytical Issues

This section provides theoretical aspects and concepts that form the basis of the argument of this chapter. It provides a synthesis of science communication, migration, urbanisation and climate change as they are discussed in literature. This section shall also describe the research methods that can be used when conducting climate change and urbanisation researches. It lays a foundation on which the review of these issues in Africa shall be evaluated in order to make recommendations for policy formulation and action.

Urbanisation is the process by which large numbers of people become permanently concentrated in relatively small areas, forming cities or urban areas (Opoko and Oluwatayo, 2014). An urban area refers to the spatial concentration of people who are working in manufacturing, tertiary and service industries (United Nations Population Fund, 2007). Urban areas have essentially characterised as being means non-agricultural, but this has been changed in recent years, as urban agriculture is now perceived as a formal and legible urban activity, which can enhance urban food security. Urbanisation is mainly driven by internal rural to urban migration, where people move from rural areas to urban areas, resulting in an increase in the number of people living in cities and a decrease of those living in rural areas (Peng *et al.* 2014). Great Britain and some European countries were the first countries to be urbanised, but this happened relatively slowly, allowing governments time to plan and provide facilities for the needs of increasing urban populations. However, there is now a new and unique phenomenon of explosive and rapid urban population growth, especially in African cities. Urban populations increased from 220 million people in 1900 to 3.2 billion people in 2005 and this increase is largely attributed to migration (Opoko and Oluwatayo, 2014).

Migration is a form of geographical or spatial motion between one geographical unit and another (Sajor, 2001). It can be external, when one moves from their home country to another, or internal. This study focuses on internal migration, which exists in various forms: rural-rural, rural-urban, urban-urban and urban-rural. Migration may also be periodic, seasonal, or long-term (UNEP

1999). It is the main reason for the rapid growth of mega-cities. People may move to the city as a result of poverty in rural communities or they may be drawn by the attractions of city life (Girard *et al.* 1996; Gugler, 1997). The normal push factors for rural people are the circumstances that make earning a living impossible, land deterioration, lack of adequate land, unequal land distribution, droughts, storms, floods and a shortage of clean water. People move to urban areas in search for non-agricultural employment and trading opportunities, which can be tapped to resuscitate declining living standards in rural areas. Nowadays, urbanisation is increasing fastest in Africa, without any significant opportunities for the new migrants (UNEP 1999). Science communication is very crucial in the formulation of priority actions towards addressing climate change and urbanisation. However, this study found that science communication in Africa is still in its infancy, mainly due to the lack of advanced technologies for collecting, assessing, evaluating and disseminating the impacts of climate change (Fiondella, 2011).

Science communication is a two-way communication, which entails the use of media, dialogue and skills in order to enhance public knowledge on the subject hence prepare it [the public] for action (Bultitude, 2011). McCallie *et al.* (2009) have defined a similar concept known as Public Engagement with Science (PES) as a process that involves scientists and the public working together and engages people with varied backgrounds and scientific expertise to articulate and contribute their perspectives, ideas, knowledge and values in response to scientific questions or science related controversies. This aspect of science communication shows that there is a multi-directional dialogue among people from different backgrounds, which allows all participants to learn on the subject at hand, in this case climate change and climate science. For science communication to be effective, it needs to be rich in content, jargon-free and should be communication based. Goh *et al.* (2008) have defined "content-rich communication" as one which is abounds with data and ideas, while "jargon-free" refers to the removal of scientific notation and technical language that scientists use to communicate with fellow scientists to maintain a common language basis with the audience in explaining concepts. Science communication should also focus on the intended audience and make sure that there is a broader base of understanding and accessibility. Different research paradigms such as the positivist, empirical and interpretivism are discussed in the subsequent paragraphs.

The positivist research paradigm has its roots in physical science and uses a systematic, scientific approach to research. Hughes (2001) explains that the positivist paradigm sees the world as being based on unchanging, universal laws and knowledge of these universal laws as explaining everything that occurs around us. Positivist research is grounded in the notion that events and phenomena around us should be observed and recorded systematically and their underlying causes understood. Blaikie (1993) postulates that positivism is based upon values of reason, truth and validity and focusing purely on facts gathered through direct observation and experience and measured empirically using quantitative methods like surveys, experiments and statistical analysis. The notion of positivism has been referred to as 'sustainability science' in scholarship treatment (Komiyama and Takeuchi 2006, Kajikawa *et al.* 2007). Chanza (2014) argues for the need for positivism to the used in seeking sustainable climate interventions for communities facing climatic opportunities and risks. This approach is aimed at developing climate resilient communities. Johnson and Christensen (2008) add that positivist research makes assumptions and is based on agreed norms and practices and the idea that it is possible to distinguish between more and less plausible claims and that science cannot provide all the answers (see Box 2.1). More information on Positivism is linked to Auguste Comte, a sociologist who coined the term "positivism".

The term "empirical" is typically associated with hypotheses that can be verified or falsified according to publicly observable behaviour (Brulle, 2015). The concept of empiricism accounts for knowledge gained and facts established through personal experience and inter-subjective relations. People can form agreements as they communicate and interact with one another. When adopting the empirical approach, no participant in dialogue can lay claim to prescriptive truth or legitimacy because it is a process of understanding a decision that must produce agreement over normative truth claims and legitimacy (Brulle *et al.* 2012). The adoption of an empirical approach that reflects what people actually value and accept when faced with the task of making decisions is critical in making ethical deliberations in this world characterised by deepening complexities and unpredictability.

Box 2.1: Positivism in Early Empiricism (Murphy, 1930; Cranston, 1957; Herrnstein and Boring, 1966; Rossi (1968; Myers, 2004)

Early empiricists and founders of modern science such as Francis Bacon believed that ideas and knowledge come from our senses and experiences (Cranston, 1957). They birthed the foundation of positivism. John Locke, an English philosopher in the 17th century supported this view and believed that with positivism, all authentic knowledge allows verification (Myers, 2004). This theory asserts that authentic knowledge assumes that the only valid knowledge is scientific. Positivism rejects innate ideas and admits that science can only study that which can be observed. Positivism is therefore characterised by quantitative approaches.

An important pillar to the positivism theory is the natural order in which the sciences must stand. Science studies what is observable, and this forms the core of positivism. Myers (2004) asserts that key features of positivism include the conviction that science is collective, that it rests on results that are dissociated from personality and social position of the researcher and that science is transcultural. Myers also believes that the science of psychology has developed through the combination of the study of philosophy and biology. Positivism has developed through the combination of the study of philosophy and biology. The ideas of philosophy, particularly empiricism have contributed to the modern theory of learning and understanding the human mind (Rossi, 1968). The only difference is that in contemporary social science, positivists are in favour of methodological debates concerning clarity, reliability and validity in greater detail.

Interpretivism research acknowledges that people's knowledge of reality is a social construction by human actors. The enquirer uses his/her preconceptions to guide the process of enquiry (Walsham, 1995). Bryman and Bell (2007) posit that interpretivism is an epistemological position that requires the social scientist to grasp the subjective meaning of social action. A common view is that the subject matter of social sciences, which is people and their institutions, is fundamentally different from that of the natural world. In keeping up with Bryman and Bell, "…the study of the social world therefore requires a different logic of research procedure, one that reflects the distinctiveness of humans against the natural order" (2007:17). Interpretivism research can be fully supported by the concept of indigenous knowledge, which puts the ideas and knowledge of local people at the forefront before making any decisions.

Indigenous knowledge is understood as "… the unique knowledge confined to a particular culture or society" (Chanza, 2014:17). Communities generate and transmit knowledge over time, so as to cope with their own agro-ecological and socio-economic environments. Indigenous knowledge is rooted from a particular place and set of experiences and generated by people living in those places. Hence, so it is only useful in its place of origin not in another place, as this will dislocate the knowledge and make it meaningless (Owusu-Ansah and Mji, 2013). Proponents of such community-based knowledge believe that local people have natural intelligence. This intelligence is used to generate knowledge through a systematic practice of observing local conditions, experimenting with solutions and readapting previously identified solutions. These can be modified situations which maybe environmental, socio-economic and technological circumstances (Brouwers, 1993; Chanza, 2014).

Results

This section highlights the extent of migration and urbanisation in Africa and how these are related to climate change. Africa is reportedly a late starter in the urbanisation race. However, it is urbanising at an alarming rate and predictions are that it will enter the urban age around 2030, when half of Africans will live in urban areas (Opoko and Oluwatayo, 2014). As illustrated in Box 2.2 which describes the urbanisation of the African city of Lagos, this trajectory is riddled with numerous challenges.

Studies on climate induced migration show that, there is a clear link between climate change disasters, displacement and migration (Penning-Rowsell *et al.* 2013), though this link is not mono-causal (Kolmannskog, 2009). Migration can be in form of displacement that is linked to sudden-onset disasters, such as the floods and storms which occurred in Malawi and Mozambique. Secondly, displacement can be linked to slow-onset disasters, such as drought (Mohamoud *et al.* 2014). In this case, people gradually move from places which can no longer produce food to those which still have a conducive climate or to urban areas in search of tertiary and manufacturing employment. A study conducted by the Norwegian Refugee Council revealed that Somalians believe that drought intensifies conflict by increasing competition over fertile land and resources (, 2009). The biggest challenge in rural settlements in Africa is the lack of profitable land, which forces poor farmers to move to cities in search of non-agricultural livelihood (UNEP 1999).

Box 2.2: The Dreading Urbanisation of Lagos, Nigeria (Lagos State Government 2006; Opoko and Oluwatayo, 2014)

Nigeria has witnessed large volume of internal migration which is induced by climate change which leads to scarcity of fertile land, impoverished soil, declining crop yields, poor harvests and soil erosion. The country is experiencing urban population growth of 4/5% which is very high when compared to the 2% experienced at the global level. The most noteworthy form of migration is movement from rural areas to urban centres which is responsible for the depopulation of some rural areas and the incursion of people into towns and cities. The migrants have mainly targeted urban centres of Lagos, Port Harcourt, Warri, Jos, Kaduna and Kano causing problems of urban transport congestion, overcrowding, poor housing, poor environmental sanitation, unemployment, crimes and other social vices which have come to characterise Nigeria's large urban centres. The country is one of the least urbanised world regions, yet experiencing the phenomenon of over-urbanisation. This is so because urban growth is not in response to industrialisation; hence, there is a high level of unemployment, low productivity, an overstuffed tertiary or service sector and marginalisation of the labour force in the towns and cities. On the other hand, Lagos serves as the economic and financial nerve-centre since it accounts for over 70% of Nigeria's industrial and commercial establishments. For this reason, Lagos has become the hub of intense settlement; and the prime destination of local and international migrants which puts pressure on land for housing and business premises and consequently leading to profound environmental implications. Lagos state has a population of 17 million with a density of 4,193 persons per sq. Km. It is estimated if the state continues to grow at the present rate of 600,000 per annum it shall become the third largest mega city in the world by 2015 following Tokyo in Japan and Bombay in India. The built-up areas of Metropolitan Lagos have an average density of over 20,000 persons per square km. Metropolitan Lagos covers 37% of the land area of Lagos State yet it is home to over 85% of the State population. There is fear that Lagos population is growing ten times faster than New York and Los Angeles with grave implication for urban sustainability.

Scholars, generally, agree about the need for better understanding of the African climate, especially in terms of its drivers and its linkages to global warming (Joubert, 2001, Chanza 2014). Even if African meteorological science has advanced considerably, there is uncertainty about major climate trends both at continental level and for individual countries (Bultitude 2011). Countries such as Tunisia, Nigeria and Namibia are signatories to United Nations Framework Convention on Climate Change (UNFCCC), which focuses mainly on medium- to long-term projections of carbon emissions and forecasting models of global warming. Namibia has established early warning systems and information centres in high-risk areas such as Ongwediva, Katima Mulilo and Mariental to make climate data more accessible. Nigeria has also designed and set up a national climate change knowledge platform that serves as a repository for data and studies to support information and data sharing, linked with the Adaptation Learning Mechanism (*ibid.*). However, these countries which are signatories to the UNFCCC lack in relevant data needed for action.

The most worrying fact is that about three-quarters of Africa's population depend on agriculture as its main source of livelihood. Since agriculture is dependent on rainfall, the change in weather patterns puts the livelihoods of huge numbers of predominately poor people at risk (Cunningham and Jacques 2006). Pastoralist areas such as Ethiopia, Kenya and the Somalia border have been severely impacted by recurrent droughts, resulting in many civil wars as people scramble for the few remaining arable land for grazing and cultivation. FAO identified the Horn of Africa (Kenya, Somalia and Ethiopia) (see Box 2.3), Zimbabwe, Malawi and Zambia as regions in Africa at most risk of drought and famine (Kirkbride and Grahn, 2008). These droughts have already been felt in Malawi, for example in 2003, where thousands of poor people living in rural areas died of hunger (Cunningham and Jacques 2006). Moreover, as populations have increased, people have been pushed out onto less productive areas of land, which are even more susceptible to drought.

An earlier report by the IPCC (2001) indicates that temperatures in Africa had risen by 0.6° C during the last century. The effects of this are two-fold: in some wet, tropical regions, rainfall is increasing, while in already arid areas, there is even less rain. By 2050, averages temperatures in Africa are predicted to increase by 1.5 to 3°C and will continue further upwards beyond this time (IPCC 2007). Warming is very likely to be greater than the global annual mean warming throughout the continent and in all seasons, with drier subtropical regions warming more than the moister tropics. The gradual yet dramatic disappearance of the glaciers on Mount Kilimanjaro is a result of climate change (UNEP 2012). The glaciers act as a water tower and several rivers are now drying up. It is estimated that 82% of the ice that capped the mountain, when it was first recorded in 1912, is now gone (*ibid.*).

Box 2.3: Drought in the "Horn of Africa" (IPCC, 2007; Kirkbride and Grahn; 2008)

Between July 2011 and mid-2012, a severe drought affected the entire East Africa region and was said to be "the worst drought in 60 years." The region is seriously affected with relentless droughts, heat stress and flooding which have led to a reduction in crop yields and livestock productivity. For this reason, the region is facing the worst food crisis in the 21st century. 12 million people in Ethiopia, Kenya and Somalia are in serious need of food. Rainfall has been below average with 2010/2011 being the driest year since 1950/1951. This causes serious problems in the region, since it is almost exclusively dependent on rain for its agriculture. The drought started in April 2011, with below normal rainfall reported in Ethiopia, Kenya and Somalia, leading to a delay of the main cropping season. Maize sowing was delayed by 10-30 days throughout the region. This drought was exacerbated by the fact that it was accompanied by the 2010 La Niña phenomena which resulted in low rainfalls. The 2011 drought prolonged the period of stress, particularly on pastoralists who depend on natural vegetation. The most worrying detail is that most parts of Somalia and Kenya as well as southern Ethiopia are expected to continue to remain dry.

A number of countries in Africa already experience considerable water stress as a result of insufficient and unreliable rainfall that changes rainfall patterns or causes flooding (see Box 2.4).

Box 2.4: The 2000/2001 Floods in Mozambique (Save the Children, 2003; Moore, 2013; UNDP, 2014)

The three worst floods recorded in Mozambique occurred first in 2000/1, the second flooded Central Mozambique in 2007/8 and the most recent one in 2013. Mozambique is extremely prone to persistent natural hazards, such as floods, tropical storms and drought. This is primarily because the country is located on the south eastern coast of Africa and is downstream from several major rivers. The extensive floods that desolated Mozambique in 2000 and 2001 killed approximately 800 people and it was recorded as the worst flood in Mozambique in 50 years. In October and November 1999, heavy rainfall affected Mozambique, followed by a period of heavy rainfall in January 2000. By the end of January 2000, the rains caused the Incomati, the Umbeluzi and the Limpopo rivers to burst their banks, flooding portions of the capital Maputo. The heavy rains in January 2000 were followed by tropical Cyclone Connie, which flooded Maputo and the southern watersheds. Subsequently, Cyclone Eline came in late February when flood waters were beginning to recede and destroyed many more homes and lives. At Chókwè, the Limpopo River reached a level 6 m (20 ft) above normal on January 24, twice its normal level. Some areas received a year's worth of rainfall in two weeks. The country lost one third of its crops, roads and railway lines were destroyed, rural settlements disappeared and hundreds of thousands of people were made homeless. Up to now in 2015, people are still living in recovery shelters with fluctuating water supplies. Roads and bridges were washed away isolating people from much needed assistance. 350,000 people lost their jobs, undermining the livelihoods of 1.5 million people. In February 2001 there was further flooding predominantly on the flood plains of the lower Zambezi River. By May 2001 around 44,000 peasant-farming families had been affected with the loss of 27,000 hectares of crops since 1,400 km² of arable land was affected. The government of Mozambique disbursed 15 million dollars to its citizens in compensation for damaged property and loss of income.

Research approaches for Urbanisation and Climate Change in Africa

Africa has challenges in simulating future projections on climate variances. Most countries do not have empirical data and where it is available, it is only about the present and the past, making it difficult to plan for the future. Lack of readily available data and data collection technology make it difficult for climatologists to make forecasts about the future state of the region or the world (Fiondella, 2011). Normal science or empirical methods are used in Africa to try to build understanding of many sub-disciplines. Studies usually try to determine the types of technologies that will exist in the future, the volume of greenhouse gases that will be emitted and their likely impacts to the climate by looking at biogeochemical cycles. There has been a sustained and systematic effort to use quantitative methods to assess the conflict potential of global climate change in Africa (Selby 2014). The positivism research paradigm is used to substantiate links between environmental and conflict variables. A case in point is that of the climate war in Darfur during 2003–2005, which Selby (2014: 837) termed as "the most celebrated modern climate war". There was evidence of credible links between drought and the region's civil war.

In Africa, indigenous knowledge is the major form of interpretivism research that is used in climate change (Bamigboye and Kuponiyi, 2010). The major characteristic of using indigenous knowledge in interpretivism, for which it is most criticized, that it is passed from generation to generation, usually by word of mouth and cultural rituals. Indigenous knowledge has been the basis for agriculture, food preparation and conservation, health care, education. It also applies and the wide range of other activities that sustain a society and its environment in many parts of the world for many centuries. Indigenous knowledge systems have a broad perspective of ecosystems and of sustainable ways of using natural resources (Owusu-Ansah and Mji, 2013). However, there is still a grave risk that much indigenous knowledge is being lost and, along with it, valuable knowledge about ways of living sustainably both ecologically and socially.

Conclusion

Africa is faced with many problems associated with climate change and urbanisation. The most worrying fact is that Africa is at more risk of climate change and is also experiencing high rates of urbanisation when compared to other regions worldwide. Despite this, it is still less developed and very incapable of conducting meaningful research that can inform local decision-making and direct action towards dealing with the two problems of climate change and urbanisation. Without proper use of research methods, results derived from studies carried out using individual methods will continue to make it difficult for climate and urbanisation studies to make valid contributions towards development of the region. This chapter concludes that use of the multi-method approach can contribute meaningful and sufficient answers to the conundrum faced by the African region. For policy and data to be useful, there is need for governments and policy-makers to sensitise the public on the issues of integrating climate knowledge into practice and policy as highlighted in Figure 2.1. – in their decision-making.

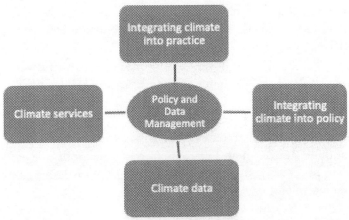

Figure 2.1.: Four pillars for successful climate data usage (Goddard *et al.* 2014)

There is need to integrate climate into policy as well as action and to invest in climate services and climate data so that all action towards addressing climate change is conversant. The continent should endeavour to integrate climate risk management into climate-sensitive development processes as an urgent and top priority requirement for the development of the region.

References

Aliyu, A. A., Bello, M. U., Kasim, R., and Martin, D. (2014). Positivist and non-positivist paradigm in social science research: Conflicting paradigms or perfect partners. *J. Mgmt. & Sustainability*, 4(3), 79-95.

Anable, J., Lane, B., and Kelay, T. (2006) *An Evidence Base Review of Public Attitudes to Climate Change and Transport Behaviour.* Research report for The Department for Transport for UK, London.

Bamigboye, E. O., & Kuponiyi, F. A. (2010). The characteristics of Indigenous Knowledge Systems (IKS) influencing their use in rice production by farmers in Ekiti State, Nigeria. *Agricultural Journal*, 5(2), 74-79.

Banda, F. (2013). *Climate Change in Africa: A Guidebook for Journalists.* United Nations Educational, Scientific and Cultural Organisation (UNESCO), Paris

Dunlap, R. E., and Brulle, R. J. (Eds.). (2015). *Climate change and society: Sociological perspectives.* Oxford University Press.

Brulle, R. J., Carmichael, J., and Jenkins, J. C. (2012). Shifting public opinion on climate change: an empirical assessment of factors influencing concern over climate change in the US, 2002–2010. *Climatic change, 114*(2), 169-188.

Bultitude, K. (2011). *The Why and How of Science Communication.* In: Rosulek, P. (Ed.). "Science Communication". Pilsen: European Commission.

Chanza, N. (2014). Indigenous knowledge and climate change: Insights from Muzarabani, Zimbabwe, PhD Thesis, Nelson Mandela Metropolitan University: Port Elizabeth.

Cranston, M. (1957). John Locke: A biography. London: Longman's Green and Co.

Cunningham, M., and Jacques, A. (Eds.). (2006). *The climate of poverty: facts, fears and hope.* Christian Aid, London

Fiondella, F. (2011). Hybrid Climate Data for East Africa. Available online: http://blogs.ei.columbia.edu/2011/09/28/hybrid-climate-data-for-east-africa/ [Accessed on 09/09/15]

Goddard, L., Baethgen, W. E., Bhojwani, H., and Robertson, A. W. (2014). The International Research Institute for Climate & Society: why, what and how. *Earth Perspectives*, *1*(1), 1-10.

Goh, B., Pomsagun, A., Le Tissier, M., Dennison, W. C., Kremer, H. H., and Weichselgartner. J. (Eds.) (2008). *Science Communication in Theory and Practice*. LOICZ Reports and Studies No. 31, 101 pages. LOICZ, Geesthacht, Germany.

Herrnstein, R., and Boring, E. (1966). *A source book in the history of psychology*. Massachusetts: Harvard University Press.

International Organisation for Migration (IOM). (2008). *Migration and Climate Change*. International Organisation for Migration (IOM), Geneva

IPCC. (2007). *Climate Change 2007: Impacts, Adaptation and Vulnerability*. Contribution of Working Group II to the Fourth Assessment Report of the Intergovernmental Panel on Climate Change. In, Parry, M. L., Canziani, O. F., Palutikof, J. P., Van der Linden, P. J., and Hanson, C. E. (Eds.). Cambridge University Press, Cambridge, 976pp.

IPCC. (2013). *Climate Change 2013: The Physical Science Basis*. Contribution of Working Group I to the Fifth Assessment Report of the Intergovernmental Panel on Climate Change. In, Stocker, T. F., Qin, D., Plattner, G. K., Tignor, M., Allen, S. K., Boschung, J. (Eds.). Cambridge University Press, Cambridge, 1535 pp.

IPCC (2014). *Climate Change 2014: Impacts, Adaptation and Vulnerability. Part A: Global and Sectoral Aspects*. Contribution of Working Group II to the Fifth Assessment Report of the Intergovernmental Panel on Climate Change. In, Field, C. B., Barros, V. R., Dokken, D. J., Mach, K. J., Mastrandrea, M. D., Bilir, T. E. (Eds.). *Cambridge* University Press, Cambridge, United Kingdom and New York, NY, USA, 1132 pp.

Joubert, M. (2001). Report: Priorities and challenges for science communication in South Africa. *Science Communication*, 22(3), 316-333.

Kajikawa, Y., Ohno, J., Takeda, Y., Matsushima, K., and Komiyama, H. (2007). Creating an academic landscape of sustainability science: an analysis of the citation network. *Sustainability Science*, 2(2), 221–231.

Kirkbride, M., and Grahn, R. (2008). *Survival of the fittest: Pastoralism and climate change in East Africa.* Oxfam International, Oxford.

Kolmannskog. V. (2009). Climate change, disaster, displacement and migration: Initial evidence from Africa Research Paper No. 180 Norwegian Refugee Council

Komiyama, H., and Takeuchi, K. (2006). Sustainability science: building a new discipline. *Sustainability Science,* 2(1)1–6.

Watson, R. T., and Albritton, D. L. (Eds.). (2001). *Climate change 2001: Synthesis report: Third assessment report of the Intergovernmental Panel on Climate Change.* Cambridge University Press.

Mohamoud, A., Kaloga, A., and Kreft, S. (2014). *Climate change, development and migration: an African Diaspora perspective.* German Watch, Bonn

Moore, M. (2013). Flooding in Mozambique, not a repeat of 2000 – 2001 Available online: https://landminesinafrica.wordpress.com/2013/01/27/flooding-in-mozambique-not-a-repeat-of-2000-2001/ [Accessed on 19/08/2015]

Morito, B. (2010). Ethics of climate change: adopting an empirical approach to moral concern. *Human Ecology Review,* 17(2),106-116.

Murphy, G. (1930). *Historical introduction to modern psychology.* New York: Harcourt, Brace, and Co.Inc.

Myers, D. (2004). Psychology (7th ed.). Michigan: Hope College.

Opoko, P. A., and Oluwatayo, A. A. (2014). Trends in urbanisation: implication for planning and low-income housing delivery in Lagos, Nigeria. *Architecture Research,* 4(1A), 15-26.

Owusu-Ansah, F. E., and Mji, G. (2013). African indigenous knowledge and research. *African Journal of Disability,* 2(1), 1-5.

Peng, X., Chen, X., and Cheng, Y. (2014). Urbanisation and its consequences. Encyclopaedia of life Support Systems (EOLSS), *Demography,* 2(1),1-12.

Penning-Rowsell, E. C., Sultana, P., and Thompson, P. M. (2013). The 'last resort'? Population movement in response to climate-related hazards in Bangladesh. *Environmental science & policy, 27,* S44-S59.

Rondinelli, D. A., Nellis J. R., and Cheema, S. (1983). *Decentralization in developing countries.* Washington, D.C, World Bank

Rossi, P. (1968). *From magic to science.* London: Routledge and Keegan Paul.

Save the Children. (2003). *Mozambique Floods Feb 2003*. Available online: http://reliefweb.int/report/mozambique/mozambique-floods-feb-2003 [Accessed on 03/09/2015]

Schneider, S. H. (2006). Climate Change: Do We Know Enough for Policy Action? *Science and Engineering Ethics*, 12 (4),607-636

Selby, J. (2014). Positivist Climate Conflict Research: A Critique. Routledge, *Geopolitics*, 19(4), 829-856

Senanayake, S. G. J. N. (2006). Indigenous knowledge as a key to sustainable development. *Journal of Agricultural Sciences–Sri Lanka*, 2(1),1-16.

UNDP. (2011). Africa Adaptation Programme: Capacity Building Experiences Improving Access, Understanding and Application of Climate Data and Information. Discussion Paper Series Vol. 2. UNDP, New York

UNDP. (2014). Recovery from Recurrent Floods 2000-2013. United Nations Development Program,

UNEP. (2012). *Thematic Focus: Climate change and Ecosystem management. Africa without Ice and Snow*. Nairobi: United Nations Environment Programme.

UNHABITAT. (2010). *The State of African Cities 2010: Governance, Inequality and Urban Land Markets*. Nairobi: UNHABITAT.

United Nations Population Fund. (2007). *State of World Population 2007: Unleashing the Potential of Urban Growth*. UNFPA, New York.

Nations, U. (2014). World urbanization prospects: The 2014 revision, highlights. department of economic and social affairs. *Population Division, United Nations, 32*.

Walsham, G. (1995). The emergence of interpretivism in IS research. *Information systems research*, 6(4), 376-394.

Chapter 3

Climate Change and Ward 16 Women in Goromonzi

Angeline T Maturure & Conillious Gwatirisa

Introduction

Climate change presents one of the biggest threats globally. Major adverse impacts of climate change include; declining water resources, water shortages and reduced agricultural productivity (IPCC, 2007). Direct health impacts from climate change include injury and death from more frequent extreme weather events such as floods and hurricanes (Alam *et al.* 2015). Dankelham (2002) opines that climate change has gender specific implications in terms of both vulnerability and adaptive capacity. The impact on women is further exacerbated by existence of structural differences between men and women through for example, gender specific roles in society. Climate change is likely to exacerbate water shortages which mostly affect women as they are largely responsible for water management in communities and in the home. Water shortages will result in greater time spent by women fetching water, hence, promoting the likelihood of disease outbreaks and further diverting women from other economic pursuits.

Undeniably, climate Change is one of the most daunting global challenges of our time (IPCC, 2007). The gravity of the situation is evidenced by that, the warmest ever recorded average global temperatures occurred between 1997 and 2011. The World Meteorological Organisation (2012) has indicated that the extent of Arctic sea ice in the year 2011 was the second lowest on record and its volume was the lowest. This and other climate and global warming effects, including thawing permafrost, sea level rise, increased flooding and shifting seasons, can in turn have negative implications for biodiversity, coastal systems, freshwater resources and livelihoods (IPCC, 2007). Over the next decades, billions of people, particularly those in the developing countries, are expected to face severe food shortages that will trigger waves of migration and displacement (UNFCCC, 2007). These migrations are often very dangerous to women and girls. In order to reach a country where they can seek asylum, many must rely on smugglers, resort to desperate measures and endure perilous routes (UNFCCC, 2007. The gravity of the impact of Climate Change is evidenced by Cyclone Nargis that struck the Irrawaddy Delta region in Myanmar in May 2008 and led to the displacement of over 800,000 people (UNFCCC, 2007).

Zimbabwe's climate is becoming warmer and drier as a result of climate change and variability (The SADC Gender Protocol Barometer on Zimbabwe, 2013). The annual mean surface temperature has warmed by about 0.40 degrees from 1900 to 2000. The timing and amount of rainfall received are becoming increasingly uncertain. The last thirty years have shown a trend towards reduced rainfall or heavy rainfall and drought occurring back to back in the same season. The frequency and length of dry spells during the rainy season have increased while the frequency of rain days has been reducing (UNFCCC, 2007). Climate change has adverse impacts on women as a result of increased periods of droughts and poor rains. The impacts on marginalized communities such as women is further exacerbated by that these communities rely on rain fed agriculture (Slingo, 2005. Since the majority of communal areas comprise women or are female-headed households they are more prone to food insecurity due to persistent dry spells (*Ibid.*). Millions of poor people faced hunger and poverty in 2015 because of droughts and erratic rains as global temperatures reach new records and because of the onset of a powerful El Niño in the Pacific (Oxfam Media Briefing, 2015). The development of the El Nino is associated with extreme weather patterns to several regions of the world. Despite the ravaging impacts of the El Nino induced droughts, external funding is not certain as Africa and the rest of the developing world continue to feed on the crumbs of the United Nations financial system that has failed to deliver on the promise of US30 billion support in fast –start finance for mitigation and adaption (IPCC, 2007).

The rise in temperature in the last century by almost 0.5 degrees Celsius lies below the global average of 0.74 degrees Celsius has serious implications on Zimbabwe as it could trigger massive crop failures and water shortages (Stinting, 2012). In the Zimbabwean context, the food and water crises are further aggravated by that about 86% of the women depend on the land for their livelihood (SADC Gender Barometer, 2013, 147). Article by Gogo in *The*

Herald (11 March,2016) posits that, a single season's lack of rain (2015/2016) left the country in need of US1, 5 billion to mitigate the impact of drought on humans, livestock and wildlife. The food and water crisis were worsened by Zimbabwe's financial woos when in 2015 revenue fell by 2, 8 percent forcing the Zimbabwean Government to look elsewhere to secure funding (Gogo, 11 March 2016).The 2015/2016 El Nino induced drought had dire consequences on the Zimbabwean economy as it is primarily agro-based with over seventy percent of the population living in rural areas and are dependent on climate sensitive livelihoods such as arable farming and livestock rearing (Zhakata, 2011).

Theoretical Perspectives

The study on climate change and its effects on women are based on various theories that act as frameworks for the understanding of the subject. The study explored the effects of climate change on women using the Sustainable Livelihoods Framework (SLF) to address the objectives of this study. Figure 3.1 illustrates the Sustainable Livelihoods Framework also known as the Sustainable Livelihoods Approach (SLA). The SLF opines that the livelihoods of an individual or a household are their means to survival. Livelihoods comprise of capabilities, assets and activities required for a person to earn a living (Chambers and Conway, 1991). The SLA posits that communities are vulnerable to seasonal shocks such as Climate Change which influence their livelihoods assets. The Sustainable Livelihoods Approach (SLA) categorizes these assets into the following five groups: human assets (including skills, knowledge, health and ability to work), social assets (comprising of informal networks, membership of formalised and relationships of trust that facilitate cooperation). There are also natural assets such as land, oil, water, soil, forests and fisheries and physical assets including basic infrastructure such as roads, water and sanitation. Lastly there are financial assets, and these are savings, credit and income (Carney, 1998). Therefore, this approach clearly explains the impact of climate change on women because it highlights that some shocks such as drought or floods affect the livelihood assets of the people (Chambers and Conway, 1991). The framework posits that floods destroy physical assets such schools, hospitals and roads and it affects women more due to their multiple

37

roles of fetching water and firewood. Figure 3.1 shows the SLA Model /SLF Framework.

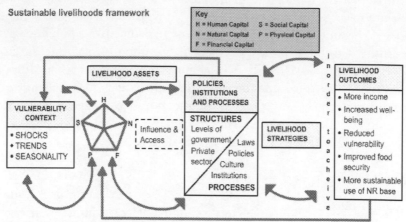

Figure 3.1: The Sustainable Livelihood Framework (Wright, 2012)

The Sustainable Livelihoods Approach also conveys that livelihood strategies are a set of activities and choices that people undertake in order to achieve their livelihood goals. (Carney, 1998). The Sustainable Livelihoods Approach posits that there are inter – related issues of social relations, social and political organisations, governance, service delivery, resource access institutions, policy and policy processes (Carney, 1998). The approach highlights the influence of policies, institutions and legislation as well as the importance of reforming livelihoods Therefore, bringing women's issues and climate change to policy debates. Livelihood outcomes are also components of the SLA and are also part of this study and these are the desired goals that people seek to achieve through their livelihood strategies (Carney, 1998). A lot of debate has raged on in the academic cycles on the reasons why women are worst affected by climate change. Confalonieri (2007) has observed that,

'Men and women are affected differently in all phases of a disaster, from exposure to risk and risk perception, to preparedness behaviour, warning communication and response. It was also noted that that there were differences in terms of physical, psychological, social and economic impacts, emergency response and ultimately to recovery and reconstruction''.

Despite the international community's increasing acknowledgement of the differential experiences and skills, women and men bring to development and environmental sustainability effort. Women still have lesser economic, political and legal clout and are hence less able to cope with and are more exposed to the adverse effects of the changing climate (Fordham, 2003). Nelleman *et al.* (2012) have observed that women in developing countries are particularly vulnerable to climate change because they are highly dependent on local natural resources for their livelihood. Women were worst affected as traditionally they were responsible for securing water, food and fuel for cooking and heating (Confalonieri, 2007).

Lack of sex disaggregated data in all sectors (e.g. livelihoods, disasters' preparedness and protection of environment, health and well-being) often lead to an underestimation of women's roles and contributions (Nelleman *et al.* 2012). This situation resulted in gender-blind climate change policy and programming, which did not take into account the gender differentiated roles of both women and men (i.e. their distinct needs, constraints and priorities (Fordham, 2003). At times, poorly crafted policies and programming may have the unintended effects of actually increasing gender-based vulnerability (Parry *et al.* 2010). At continental level impacts of climate change were being felt by people across Africa as evidenced by the change in temperatures. These increased temperature changes affected the health, livelihoods, food productivity, water availability and overall security of the African people (Parry *et al.* 2010). Since women comprised the majority of the global agricultural workforce in developing countries, they were obliged to adapt to increased instances of drought and desertification (Cutter, 1995).

To illustrate this, in the 2011 Kenyan drought, men migrated away from rural communities with their livestock in search of water and pastures, while women were left in charge of households with very few resources. This led to an increase in petty trade and sex work, which also increased their risk of contracting HIV and AIDS (Cutter, 1995). In the Kenyan experience, women were not allowed to make the decision to sell or slaughter livestock without the permission and supervision of men (*Ibid.*) in these pastoral communities; women had to wait for men to return hence exposing them to the precarious food insecurities.

Persistent droughts in Zimbabwe have severely strained surface and ground water systems, contributing to the country's deteriorating water supply (Chaguta, 2010). The lack of access to clean drinking water also disproportionately impacted on women. In many communities around the world, where dependable irrigation is a distant dream and clean water a precious commodity, women and girls bear the primary burden of finding water (*Ibid.*). To illustrate climate change impacts on women more than men, it is estimated that globally, women and children collectively spend 140 million hours per day collecting water for their families and communities, resulting in lost productive potential (UNICEF, 2008). This is time not spent working or engaging in income-generating jobs, or caring for family members, or attending school (*Ibid.*). In Sub-Saharan Africa, women and girls collectively spend a total of 40 billion hours per year collecting water for their households. Travelling long distances to search for water, especially in remote areas, also increases the risk of sexual violence for women and girls (UNICEF, 2008). This is particularly true in countries marred by violent conflict, such as South Sudan and Democratic Republic of the Congo, where instances of rape and abductions during water-fetching trips have been documented for years (*Ibid.*). Facilitating better access to clean water may not only help reduce the incidence of rape and abduction, but also helps fulfil the productive potential of women through increased educational attainment and economic participation. The absence of a gender policy framework on the sound management and protection of the environment and natural resources in Zimbabwe has exacerbated the impact on women (Chaguta, 2010).

Climate Change has negative health impacts to communities with sensitive diseases prevalent in poor countries that have minimal resources to treat and prevent illness. Examples of climate change related health impacts include: frequent and severe heat stress linked to sustained increases in temperature (UNICEF, 2008). The reduction in air quality that often accompanies a heat wave can lead to breathing problems and worsen respiratory diseases. The impacts of climate change on agriculture and other food systems have also increased the rates of malnutrition (UNICEF,2008). Poor women are more likely to bear the brunt of these types of health problems due to their limited access to health facilities, low awareness of risks and social and cultural norms that make women primary caregivers for family members.

Growing evidence suggests that climate change will affect human health due to increased floods, storms, fires and droughts (UNICEF, 2008). Climate changes is associated with infectious disease vectors, including the geographical range of malaria and other mosquito-borne diseases, such as dengue; increases in the burden of diarrhoea diseases and of water-borne pathogens such as cholera; and an increase in cardio-respiratory morbidity and mortality associated with ground level ozone (GOZ, 2010). Climate change is also expected to exacerbate the effects of human-induced ozone depletion in the Southern hemisphere, further worsening this situation (Karoly, 2003).

The erratic water supply situation in Zimbabwe has contributed to an increase in the outbreak of water-borne diseases (UNICEF, 2008). The gravity of impacts of Climate Change is evidenced by the outbreak of the deadly 2008 cholera outbreak that was one of the largest outbreaks in recorded history, affecting over 100,000 people and killing over 4,000 (GoZ, 2010). Moreover, the potential for cross-contamination of water and sanitation systems results in the recurrent outbreaks of cholera during the rainy season a major risk factor during flood events, as experienced in Malawi (UNICEF, 2008). Zimbabwe is also vulnerable to the high prevalence of malaria (Chigwada, 2009). According to the IPCC (2007), by 2100; changes in temperature and precipitation are likely to alter the geographical distribution of malaria in Zimbabwe, with previous unsuitable areas of dense human population becoming suitable for the transmission.

One of the major impacts of climate change is that it results in massive displacement of communities resulting in serious impacts on the political, economic and social fabric of communities (Huelsnbeck, 2012). At times women may also spend years in displacement, living in camps, integrating into urban areas or working in remote areas (Cutter, 2013). A study by Huelsnbeck (2012) established that people displaced by climate change-related events may have no way to return to their place of origin.

One of the major impacts of Climate Change is that women and girls are often attacked as they venture outside for water or firewood. Climate Change induced droughts often force adolescents to marry at increasingly younger ages (Huelsnbeck, 2012), While the displaced women may trade sex for money or precious resources may be the only way women are able to support themselves and their families (Cutter, 2013). Similarly, in Mexico desertification was more pronounced and distressing that it led to the emigration of 600,000 to 700,000 annually. The migratory consequences of environmental factors resulted in higher death rates for women in

41

least developed countries, as a direct link to their socio-economic status, to behavioural restrictions and poor access to information (Alem *et al.* 2015). Save the Children India reports that while human trafficking has always been a problem, Cyclone Aila, which displaced more than a million people in May 2009, catalysed trafficking.

As a panacea for climate change mitigation, the parties to the Cancun Agreement issued a decision that endorsed both the Convention on the Elimination of All Forms of Discrimination against Women (CEDAW) and the 1995 Beijing Declaration and Platform for Action (Alam, 2015). These parties specifically recognised the need for women's participation in the decision-making process to effectively combat, mitigate and adapt to climate change. This decision established gender as a permanent agenda item for all future COPs. It also spurred the creation of a gender-sensitive climate policy. It did not mandate the inclusion of women, but rather invited parties to meet gender-balanced goals. Additionally, the decision established a mechanism to track, for the first time in the history of COP, women's participation in the constituted bodies of COP (Alam, 2015).

Methodology

This research study utilised the Mixed Methods which embraced qualitative and quantitative data collection methods in order to complement each other. The Mixed Methods were applied for the purpose of credibility and validity of this research (Creswell, 2007). For the purpose of understanding the impacts of climate change and the copying strategies in Goromonzi Ward 16 the study used the questionnaire method so as to integrate qualitative and quantitative techniques. Figure 3.2 illustrates the study areas of two villages in Goromonzi District.

GOROMONZI DISTRICT MAP

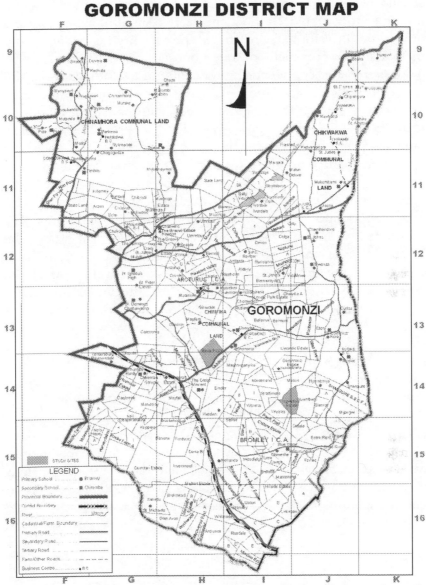

Figure 3.2: Location of the study area's two villages (Surveyor General's Office, 2004)

The study distributed 50 self-administered questionnaires to the two villages in the ward. In this study, a total of fifty questionnaires were administered through hand delivery to assess the impact of climate change on women. Forty women and ten men who reside in Ward 16 of Goromonzi were given questionnaires. Of the fifty questionnaires forty were correctly filled in and returned giving a response rate of 80 %. Respondents were assured of anonymity and privacy and therefore they felt free to provide honest responses and there was no possibility of questionnaire bias (Greener, 2008). Anonymity and confidentially was built into the questionnaire Therefore, providing respondents with an honest sense of security. Systematic sampling was used to select questionnaire respondents with the study approaching every third household. Systematic sampling was used so as to eliminate any chances of bias. The study was carried out with the assistance of a locally based research assistant. The recruitment of the local research assistance was carried out to exploit his vast knowledge of the area as well as to gain rapport into the area. The study gathered data from 8 key informants comprising the Environmental Management officer, the District Administrator, two Agritex workers, two Forestry officials and two village heads. Most of the interviews were held at the homes of the informants and offices of the informants. Assurance was given to the key informants of the confidentiality of information. The study also used non-participatory observations characterised by the observation of the fields, the available water resources, major asserts and the environment of Ward 16. The study used public records, the media, private papers, visual documents and strategies, policies, action plans among other documents. The major advantage of document analysis is, it enriches the study by giving detailed information from various sources (Bryman, 2001) Frequency tables, pie charts; bar graphs were used to present data. For qualitative data, narrative analysis was done. Data analysis is the process of developing answers to questions through the examination and interpretation of data (Kothari, 2004).

Results

The sample comprised twenty males and eighty females for comparative purposes of the service provision status by gender. Regarding their marital status, 33% claimed that they were married, twenty-eight (28%) were widowed while 22 % were single. The remainder indicated that they had re-married.

Table 3.1: Socio-demographic features (Field Survey, 2016)

A. Demographic Data of Respondents	Percentage (%) Responses
Male	20
Females	80
Marital status	
Married	33
Widowed	28
Single	22
Remarried	17

The Water Supply Situation

Most respondents indicated that boreholes were the major source of water in the study area. Basing on questionnaire responses and from the information gathered from the District Administrator's office, women in Goromonzi were facing a major challenge in water supply. They only relied on borehole water as the Chinyika River which was their only river in Ward 16 had almost dried up. One borehole is indicated to have served about 130 households with women in this area waking up as early as 3am. In most cases, women formed long queues at the boreholes forcing them to spend over five hours fetching water. Due to drought the women complained that the water table for boreholes had fallen hence forcing them to take about twenty minutes waiting for water to come. On average the women indicated that it took about 30 minutes to fill a mere 20 litre bucket. This observation concurs with Alam *et al.* (2015)'s assertion on climate change's significant effects on water sources with women bearing the burden of fetching water for their families and spending significant amounts of their time daily drawing water from distant sources.

Responses by the majority of the women indicated that the area had been severely hit by drought.

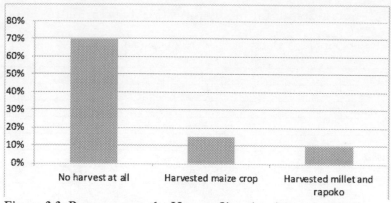

Figure 3.3: Responses on the Harvest Situation (Maturure, 2016)

Responses on the harvest situation showed that the majority 70 % had no harvest at all while 15 % only harvested the maize crop. Ten percent (10%) of the respondents stated that they had only harvested sorghum and groundnuts. Asked about who was mostly affected by climate change, 60% 0f the respondents strongly agreed that indeed women were more affected while 20% merely agreed. Fifteen (15%) were not very sure about who was more affected.

Before the drought that has ravaged Ward 16, the women outlined that they had gardens where they did horticulture and produced tomatoes, rape, covo, carrots, onions among other produce for sale in Harare. The El Nino induced drought had negative impacts on livelihoods as communities were affected by loss of income due to lack of water to water their gardens. To illustrate the gravity of the climate change menace, some women indicated that previously their area was a semi wetland area capable of retaining moisture almost throughout. This physical benefit had allowed them to grow maize twice a year but, due to climate change the wetlands had dried up. The drying up of these wetlands had reduced the women's income because they were no longer capable of growing maize known as "*mabvachando*" in Shona for 'sale when not in season'. The respondents stated that they used to harvested pumpkins in July which they grew with the moisture still on the ground. The questionnaire responses revealed that the climatic shift had caused untold suffering because they could not afford to buy food, or take their children to school. Plate 1 shows the great devastation associated with the El Nino induced drought.

Plate 3.1: Wilting maize crops observed in Goromonzi (Fieldwork, 2016)

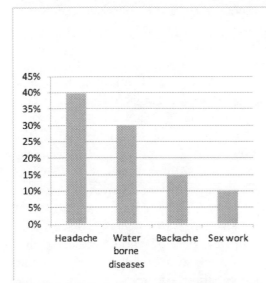

Figure 3.4: Responses on health Impacts (Maturure, 2016)

The study showed that 40% of the women suffered from headaches due to carrying buckets of water on their heads for long distances. Thirty percent (30%) stated that due to water challenges at times they had been affected by diarrhoea or dysentery. Fifteen percent (15%) indicated that they had developed backaches and

10% had stress related illnesses. This concurs with Masika (2002)'s observation that women are faced with more immediate health risks than men due to their role in the gendered division of labour.

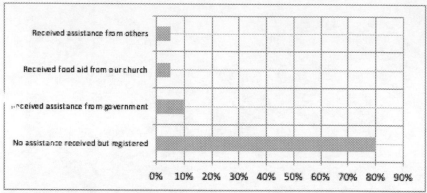

Figure 3.5: Type of Assistance received (Maturure, 2016)

At the time of the study, it was evident that 80% of the respondents had not received any form of assistance. 10% indicated that they had received assistance from the Department of Social Welfare with food handouts being only given to the terminally ill, old people, child headed families, widows and to those with more than four grandchildren and the disabled. Another Five (5%) indicated that they have received assistance from their churches. The churches that had given assistance so far were the Latter-Day Saints and Apostolic Faith Mission (AFM) and these had only assisted their congregants. The Councillor confirmed that the Ministry of Women' Affairs, Gender and Community Development had given mealie meal to other wards around Goromonzi but had excluded the two villagers. The Councillor confirmed that at the time of this study none had received. However, although 80% of the respondents indicated that their names had been written down.

With regards to coping strategies, 30% of women indicated that they had resorted to selling livestock and other properties, while 20% indicated that they had diversified into cross border trading. Fifteen percent (15%) stated that they were getting external assistance, 10% were into sex work and another 10% were into Part-Time jobs. Another survival strategy included engaging in anti-social behaviours such as in engaging in sex for money. The Chinyika-Johane Marange Apostolic sect members indicated that they were very comfortable with marrying off younger girls to older men Therefore, acknowledging Brown (2013)'s observation that as resources of families dwindle, adolescent girls are forced to marry at increasingly younger ages.

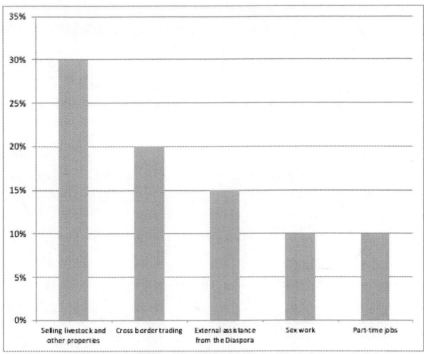

Figure 3.6: Copying strategies (Maturure, 2016)

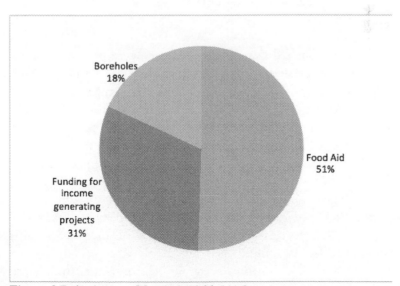

Figure 3.7: Assistance Needed (Field, 2016)

Asked about their needs, fifty percent (50%) of the respondents indicated that they wanted food aid, 20% indicated that they wanted income generating projects while- 20% stated that they needed more boreholes.

Discussion

The study established that most women were hard hit by climate change induced drought as they harvested low grain yields. These food shortages were a recipe for mass starvation and needed urgent attention. Besides this, the study area was hard hit by critical water shortages which forced women to travel long distances in search of water or to wake up as early as 3am in search of water. As a result of dwindling water sources, most community gardens were no longer productive. The responses by 80% of the women indicated that despite poor harvests, the area had not received any assistance from the government and development partners and probably this was due to their weak political capital. To reduce vulnerability, some women resorted to the selling of livestock and cross border trading. The study established that to offset the impacts of climate change induced drought some women resorted to anti-social behaviour such engaging in sex work or forcing their daughters into early marriages. The indulgence into activities such as sex work increased the vulnerability of women and girls posing a serious health hazard. The study also established that the area had very few income generating projects and had very little government assistance.

The study established that indeed women were greatly affected by the changing weather patterns in their area. There was evidence of reduced and erratic rainfall and also high temperatures well above those experienced in all the other years. The study established that climate change had resulted in serious water shortages which were further aggravated by the existence of few boreholes. The critical water shortages were further exacerbated by that most of the traditional sources of water had dried up. In addition to this, since the women previously survived on horticulture, they could no longer practice it because of water shortages and therefore, cutting their source of income. Health problems were also imminent as they drank water from unsafe sources. Questionnaire responses by women indicate that there were outbreaks of dysentery and diarrhoea. As a result of water shortages, women were forced to walk long distances in search of water. Most women also indicated that headaches and backaches were affecting women in Ward 16 of Goromonzi because of heavy workloads of fetching firewood and water.

Despite the fact that women of Goromonzi Ward 16 were facing the harsh impacts of the climate change, they received very little assistance from government and donors. In cases of crisis, women devised coping strategies which in some cases were against the social norms as evidenced by the engagement by some women into sex work and some families marrying off very young girls. This increased women's vulnerability to diseases such as Sexually transmitted diseases and HIV and AIDS.

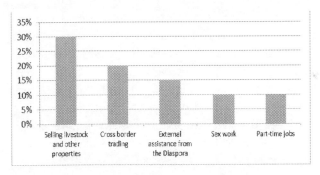

Figure: 3.8: Responses on coping strategies (Field Survey, 2016)

As a panacea to climate change mitigation some women had ventured into cross border trading while others sourced for temporary jobs in mines and farms around Goromonzi were they got meagre wages? Most women indicated that the area was faced by the critical food shortages as communities had harvested very low maize yields. The low maize harvests were a recipe for mass starvation especially with reference to women and children. The food crisis was further aggravated by the drying up of water sources as this affected all horticultural projects. The study area was therefore in dire need of cash transfers as most women indicated that they were receiving very little assistance from the Government. Women in ward 16 also indicated that they were stressed due to the hunger.

Conclusion, Policy Options and Practical Recommendations

Goromonzi Ward 16 residents were facing severe food and water shortages. As a result of the acute food shortages, some women came up with a variety of coping strategies that included selling their livestock, resorting to cross border trading and engaging in sex work. The critical food shortages were further worsened by that most households had not received any food aid from government and other development partners. The acute water shortages were evidenced by the drying up of major water sources such as the Chinyika River. In addition to this, most bore holes had dried up resulting in women waking up as early as 3 am. To mitigate the impacts of climate change, the government and donors need to introduce poverty alleviation programmes such as food for work programmes or food aid. The government and development partners also need to rehabilitate and sink more boreholes. In addition to this, the government needs to capacity build communities on irrigation farming as well as to construct small dams. Moreover, the communities need to diversify into short-season crops. In order to alleviate the impacts of the persistent droughts there is need for diversification into short-season crops and into conservation farming. Lastly, the government and development partners needed to empower communities by introducing small irrigation projects or by constructing small dams.

An evaluation of the results, based on the case study provided the study, the following recommendations are made:
- The Government should empower women with climate change mitigation strategies.
- The government and development partners need to capacity build
- communities (especially women) on soil and water conservation techniques
- The Ministry of Agriculture needs to empower communities on climate resilient open pollinated seed varieties through seed multiplication of drought, disease and pest tolerant crops.
- The government should introduce gender sensitive policies and programmes targeting climate change mitigation with the focus being to address the power dynamics between men and women at village, district and national levels.

- Farmers need to diversify into conservation agriculture and short-season crops.

References

Alam, M., Bhatia, R., Mawby, B., and Georgetown Institute for Women, Peace and Security. (2015). *Women and Climate Change. Impact and Agency in Human Rights, Security, and Economic Development*. Georgetown Institute for Women, Peace and Security.

Bartlett, S. (2001). *Key Issues for Education Study's*. London. SAGE.

Boote, D. N., and Beile, P. (2005). Scholars before researchers: On the centrality of the dissertation literature review in research preparation. *Educational researcher, 34*(6), 3-15.

Bradshaw, S. (2004). Socio- Economic Impacts of Natural Disasters: A Gender Analysis: A UN Economic Commission for Latin America and the Caribbean.

Bryman, A. (2001). *Social Research Methods*. Oxford. Oxford University Press.

Burns, R.B. (1997) Introduction to Research Methods 4th Edition. London. SAGE.

Canhao, M. T., and Keogh, E. (2000). *Research Methods: Basic Statistics Module 1*. Harare. UCDE of Zimbabwe.

Carney, D. (Ed.). (1998). *Sustainable Rural Livelihoods: What Contribution can we make?* London. DFID

Cohen, L., Manion, L., and Morrison, K. R. B. (2007). *Research Methods in Education, 6th Edition*. London, Routledge.

Confalonieri, U. (Ed.) (2007). Climate change 2007: Impacts, Adaptation and Vulnerability. Contribution of Working Group 11 to the 4th Assessment Report of the Intergovernmental Panel on Climate Change. Cambridge. Cambridge University Press.

Cooper, D. R., and Schindler, P. S. (2006). *Business Research Method, 9th Edition*. Boston. McGraw- Hill Irwin.

Cutter, S. L. (1995). The forgotten casualties: women, children, and environmental change. *Global environmental change*, 5(3), 181-194.

Dankelham, S. (2004). *Climate Change: Learning from Gender analysis and Women's Experiences of Organising for Sustainable Development*. London. Gender and Development Links.

Davison, J. (1998). *Agriculture, Women and Land: The African Experience*. Westview. Boulder and Colorado.

De Vaus, D. A. (2001). *Research Design in Social Research*. London. SAGE.

Fordham, M. (2003). *Gender, Disaster and Development: The Necessity of integration. Natural disasters and Development in a Globalising World*. London. Routledge.

Fujikura, R., and Kawanishi, M. (2012). *Climate change adaptation and international development: making development cooperation more effective*. Washington: Earthscan Routledge.

Habtezion, S. (2013). *Overview of linkages between Gender and Climate*. New York. UNDP Publishers.

IPCC (2007) Adaption and Vulnerability Summary for Policymakers. Geneva. IPCC

Johnson, S. J. (2015). *Why Climate Change is a Gender Equality Issue*. Energy Desk. Green Peace

Kothari, C. R. (2004). *Research Methodology: Methods and Techniques. 2nd Ed*. New Delhi: New Age International.

Kumar, R. (2005). *Research Methodology: A Step by Step Guide for Beginners*. New Delhi. SAGE Publishers.

Leedy, P. (1997). *Practical Research: Planning and Design. 6th Edition*. Merril. New Jersey.

Made, P. (2013). SADC Gender Protocol Barometer of Zimbabwe. Johannesburg. Gender Links.

Macmillan, J. H., and Schumacher, B. A. (1997). *Research Methods in Education*. New York Hamper and Collins Publishers.

Masika, R. (Ed.). (2002). *Gender, development, and climate change*. Oxford, Oxfam.

Monette, D. (1986). *Social Research. London*. SAGE Publications.

Nelleman, C., Verma, R., and Hislop, L. (Eds.). (2012). Women at the Frontline of Climate change: Gender Risks and Hopes –A rapid response Assessment. UNDP Publishers. New York

Parry, M. L., Canziani, O. F., Palutikof ,J. P., Van der Linden, P. J., and Harrison, C. E.(Eds.). (2010). Contribution of Working Group 11 to the 4th Assessment Report of the Intergovernmental Panel on Climate Change. Cambridge: Cambridge University Press.

Robson, C. (2011). *Real World Research. 3rd Edition*. Chichester: Willey and Sons.

Tuckman, B. W. (1972). *Conducting Educational Research*. New York. Rowman and Littlefield Publishers.

Saunders, M., Taylor, A. F., Shingle, D. L., Cimbala, J. M., Zhou, Z., and Donahue, H. J. (2007). *Research Methods for Business: A skill Building Approach, 5th Edition.* Chichester: Wiley and Sons.

Slingo, J. M. (2005). *Introduction: Food Crops in a changing Climate.* Washington. Jan Publications.

Stiftung, K. A. (2015). *Climate Change in Zimbabwe: Information and Adaptation.* Harare. Country Report Publishers.

Gogo, J. (11 March 2016). Global climate talks: Africa shoots self in foot | *The Herald,* Available online: https://www.herald.co.zw/global-climate-talks-africa-shoots-self-in-foot/ [Accessed on 17 February 2017]

Trochim, S., and William, M. K. (2006). *Research Methods Knowledge Base.* London. Sage.

UNFCC. (2007). *Climate Change Impacts, Vulnerabilities and Adaptation in Developing Countries.* San Francisco. UNFCC.

Walliman, N. (2009). *Social Research Methods.* London. Sage

Zhakata, W. (Ed) (2004). Climate Change Mitigation Studies in Zimbabwe. Harare. Government of the Republic of Zimbabwe.

Chapter 4

Urban Environmental Planning and Disaster Preparedness: A Special Focus on Beitbridge Town

Innocent Chirisa, Sharon Marimira & Artwell Nyirenda

Introduction

Climate change is now becoming a reality that many African countries have to face. There has been an increase in the occurrence of natural disasters and extreme weather conditions. Disasters which includes Floods, droughts and famines, hurricanes, pestilence and major storms, which have resulted in economic, social and physical losses, destroying development gains and exacerbating poverty. Ferris and Petz (2013) argue that high-profile, large-scale disasters are increasing universal awareness about the need to fortify national and regional capacities to mitigate, respond to and manage such events. For this reason, there is need to create resilient cities that have the capacity to resist, adapt and effectively recover from these effects. These cities are able to engage in disaster response and management and plan effectively for their environments to preserve human life, the ecology and the economy. This chapter seeks to uncover the level of preparedness and planning of Beitbridge Town regards the shocks and stresses of natural disasters and its ability to bounce back to normality. In Zimbabwe, the principal aim of planning, as enshrined in the preamble of Zimbabwe's Regional Town and Country Planning Act (Chapter 29:12), is to improve the physical, economic and social environment and in particular to promote health, safety, order, amenity and general welfare (RTCP Act: Preamble). Since 1996, the frequency of natural disasters especially floods and drought all over Zimbabwe has increased, resulting in human and socio-economic loses. Zimbabwe has great potential for growth but due to continuous loss of infrastructure, this potential has failed to be realised. This has hindered investment within small urban areas such as Beitbridge, which have great potential for growth (Kadirire, 16/04/15). Although Beitbridge, being a low-lying area, is known to be a flood prone area, there seems to be little planning or preparedness to such an eventuality, as planning authorities are more reactive than proactive.

This chapter seeks to explore the gaps and constraints and foster an understanding to enhance the environmental planning and preparedness of the Civil Protection Unit in managing floods where Beitbridge is concerned and to enhance the capacity of local authorities to effectively respond to disasters. The specific objectives are Therefore, to examine the current intervention strategies and their effectiveness; to identify challenges in implementing them; to document and discuss the views of the public towards current intervention strategies; and to recommend possible new intervention strategies. In terms of organisation, the first part of the chapter introduces the study and the scope of the research in terms of the study aim and objectives and the problem statement. The second part reviews the literature as well as concepts that have been promulgated in relation to the study. The third part focuses on the research methods used to conduct the research and the fourth part present and analyse the data gathered. The last part will analyse the findings of the study and give conclusion and recommendations.

Context of the Study

The beginning of the 20th century experienced an increase in losses from natural disaster. Due to the continuous upward trend in the present century owing to climate change, it has increased the incidence of disasters, especially meteorological ones such as floods (Ejeta *et al.* 2015). On a global scale, the need to prevent disaster led to the establishment of the United Nations World Conference on Disaster Reduction (WCRD) in Kobe, Japan 2005. It produced the Hyogo Framework for Action (HFA) for the period of 2005 to 2015. UNISDR (2012: 7) observes how the City of Kobe, Japan, with 1.5 million inhabitants, suffered great losses during the Great Hanshin-Awaji Earthquake of January 1985, disrupting the activities of its busiest ports. The HFA outcome is to ensure the substantial reduction of disaster losses in the lives and social, economic and environmental assets of communities and countries. WCDR (2005) argues, the strategy seeks to ensure national and local prioritisation of disaster risk reduction, with a strong institutional basis for implementation; identification, assessment and monitoring of disaster risks thereby enhancing nearly warnings. It also seeks to use

58

knowledge, innovation and education to build a culture of safety and resilience at all levels. In 2010, the Framework of Resilient Cities was brought forward by the United Nations Office for Disaster Risk Reduction (UNISDR) for 2010-2015, providing guidelines as to how to address disaster risk reduction and identify strategic areas for intervention.

At regional level, the Southern African Development Community (SADC) has faced a number of natural disasters, mainly floods and cyclones. In 1984, torrential rains from Cyclone Demoina reached 600 mm in 24 hours at St Lucia causing extreme floods in north-eastern South Africa, Mozambique and Swaziland, which collapsed bridges, roads and houses, causing about 2.7 million dollars' worth of damage (SADC, 2011). Cyclone Eline hit Mozambique, South Africa, Botswana and Zimbabwe in February 2000, which in Mozambique alone killed more than 700 people, displacing more than a million people and destroyed infrastructure worth $1 billion (Gwimbi, 2009). Because of these vulnerabilities, SADC created the Regional Platform for Disaster Risk Reduction, a regional cooperation strategy that countries within the community have to adhere to within their national policies. One of the policies is that member states should put in place early warning systems and resilience measures (SADC, 2015). Within the same framework, the Zambezi Watercourse Commission (ZAMCOM) was established to inform countries on the Zambezi River Basin about flood water movement downstream, meteorological data provision across the region. This will be done through water observation and information systems and track the development and extent of floods over time. This is because the Limpopo River basin is shared by four countries; the river follows the boundary between Botswana and South Africa, then between Zimbabwe and South Africa and then reaches the Indian Ocean by crossing southern Mozambique. As such, any basin management decision taken by the three upstream countries will affect Mozambique, which is located downstream (Spaliviero *et al.* 2011: 840). This illustrated in Figure 4.1.

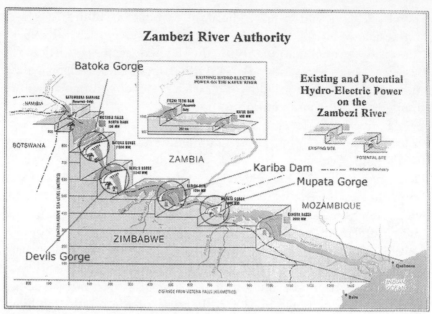

Figure 4.1: Limpopo River Basin (SADC, 2011)

In January 2011, the Zambezi River Authority informed the Mozambique authorities that Lake Kariba and Cabora Basa would be opening two of their spilling gates due to maximum water levels being reached. This was going increase water discharge toto more than 3000 m³/s, which may cause flood downstream, enabling them to take pre-emptive action in advance (SADC, 2011). Other measures that SADC members are mandated to take include mapping of areas deemed to be flood risk areas; stockpiling of relief items in districts likely to be affected by flooding; clearing of waterways in urban areas; and strengthening of transport infrastructure in critical locations (SADC, 2011).

Floods have been a common phenomenon in Zimbabwe over the last 100 years and occur every year (Zimbabwe National Contingency Plan, 2012-2013). In 2000, Cyclone Eline-induced floods in the Zambezi Basin left 90 people dead, affecting over 250,000 people and causing approximately US$7.5 million in economic losses. Floods mostly occur in the southern and northern low-lying areas of Zimbabwe, in between river confluences and downstream of major dams, which include Middle Sabi, Muzarabani, Tsholotsho, Beitbridge, Bubi and Gokwe. The most affected population groups include the women and children, single headed and child headed families. Floods also affect homes, roads, telephone and electricity supply equipment, agricultural assets and crops (Zimbabwe National Contingency Plan, 2012-2013). In February 2014, non-stop rains resulted in floods in Tokwe-Mukosi catchment area, affecting more 60 000 people and resulting in the loss of property. The Government of Zimbabwe estimated that US$20 million was needed to evacuate the affected families (Government Gazette, 2014; OCHA Report, 2014). According to a Red Cross Report (2014) by 9 February, only 36 of the targeted 2 230 families had been moved, bringing to the fore the question of the level of preparedness of the government.

Offices in Zimbabwe responsible for early warning systems are the Meteorological service Department, the Civil Protection Department/Unit (CPU) and the Zimbabwe National Water Authority (ZINWA). ZINWA monitors river flows and the state of hydrology, while the Meteorological Services Department forecasts and predicts weather conditions and the Civil Protection Unit (CPU) coordinates preparedness and response. Other partners engaged include NGOs, which provide financial, material and logistical support to the CPU to prepare and respond to floods (Zimbabwe National Contingency Plan, 2012-2013).

The new Constitution of Zimbabwe (2013) under Chapter Four (Declaration of Rights) Section 73, stipulates that every person has the right "to an environment that is not harmful to their health or well-being" and "to have an environment protected for the profit of the current and future generations, through legislative and other measures". Such measures should "prevent pollution and ecological degradation", "promote conservation" and "secure ecologically sustainable development and use of natural resources, while promoting socioeconomic development". Therefore, the new constitution calls for a safe and sustainable environment for people to live in. This implies there is need to ensure environmental planning and disaster preparedness to protect the people and their assets when the disasters occur.

The ZIMASSET (2013-2018) document observes how susceptible Zimbabwe is to perennial floods and droughts caused by climatic changes emanating from global warming which in turn affects the country's agro-based economy and communities whose livelihoods largely depend on rain fed agriculture and livestock production. To achieve this agenda, the government proposed a cluster on "Environmental Management", whose outcome is improved natural resources management. In this context, the

Ministry of Environment is mandated to continue advocacy and awareness campaigns; formulate national climate change policy and strengthen the climate and disaster management policy.

The Civil Protection policy provides that "every resident of Zimbabwe should assist where possible to avert or limit the effects of a disaster". The Department of Civil Protection, under the Ministry of Local Government Rural and Urban Development, has been given the coordinative role empowered by the Civil Protection Act of 2001 (Chapter 10:06) to establish, coordinate and direct the activities of public and private emergency services and to take maximum use of resources for disaster mitigation and set up emergency plans at all levels of government. Its primary mandate is preparing for prevention, where possible and mitigating the effects of disasters as they occur and act as a centre for the dissemination of disaster-related information. It is also there to provide humanitarian measures aimed at protecting populations, their environment and natural inheritance against accidents and disasters of every kind. To empower the populations, a National Civil Protection Fund was established to civil protection fund activities throughout the country such as research, training and acquisition of necessary property to promote the objects of the fund.

Structurally, the civil protection system in Zimbabwe has the President as Head of State, the parliament for legislation, the cabinet for policy formulation and the civil protection department for implementation. The Department of Civil Protection is decentralised at national, provincial and district levels of government to ensure the effective management of disasters and enable the preparation of civil protection plans at these levels. The department is mandated, furthermore, to work with other line ministries in the government, private and NGO organisations.

The Meteorological Services Bill (2003) states that one of the services of the Meteorological Department is to issue out weather and climate forecasts and advance warnings on weather conditions likely to endanger life and property. The Water Act (1998) law promotes Integrated Water Resources Management between the seven different catchments in Zimbabwe. Although it makes references to the management of water resources it does not make reference to flood management, indicating a weakness within the Act itself.

Theoretical Perspectives

Cities contain human agents, critical movements and spaces of change and are therefore important economic forces not only for themselves but for the entire nation (UNHABITAT, 2012). They prioritise investment in social safety nets and in local and regional infrastructure development to ensure long term growth that can stimulate consumption and ultimately growth, productivity, competitiveness and prosperity (UNHABITAT, 2012). Because of this, there is need for cities to be able to live on and prosper in the face of numerous pressures that may pose a threat to their existence. For example, the loss of infrastructure such as roads, water and communication facilities will cause cities to be less competitive at national, regional and global level, thereby restricting their growth.

Many scholars (e.g. Schipper, 2007; Verner, 2010; Tyler and Moench, 2012) argue that predicting climate risks such as natural disasters and adapting to them has been made more complex and difficult due to climate change and institutional constraints. This has resulted in the need to consider the problem as one of building resilience so that those cities are able to survive even in situations when disasters hit without warning (uncertainty). According to IPCC (2007), resilience is the capacity of a social structure to absorb disturbances while retaining the same basic structure and ways of operating, the capacity for self-organisation and to adapt to stress and change. A resilient city is therefore one which is prepared to absorb and recover from any shock or stress while maintaining its essential functions, structures and identity and adapting and thriving in the face of continual change.

Creating resilient cities goes hand in hand with creating sustainable livelihoods, as the framework considers the livelihood context in terms of vulnerability context, assets, policies and institutional strategies designed to address the problems at the local level (Spalivioro *et al.* 2011; Knutsson, 2006). Chambers and Conway (1998) define a livelihood as sustainable if it can survive and recuperate from pressures and tremors and maintain or enhance its capabilities and assets both in the current and future generations, without undermining the natural resource base. Therefore, the framework maintains that there is need to create adaptive strategies, participation and empowerment of individuals and communities to deal with vulnerabilities like disasters (Gwimbi,

2009; Chambers and Conway, 1998; Scoones, 2000). Therefore, understanding the livelihood activities of people's assets and rights of the community make it possible to create livelihoods that are sustainable and productive as well as resilient cities (Spalivioro *et al.* 2011).

There are two ways to deal with floods - structural and non-structural measures. The former comprises of structures or hard engineered infrastructure that modify floods such as hydraulic parameters that are manipulated to reduce the flood, by regulating the volume of runoff and peak discharge (Muianga, 2004: 2). The latter are preventive measures in reducing populations' vulnerability to floods. These include disaster preparedness, land use planning and flood forecasting and profiling and do not require large investment as do structural measures, but rather rely on a good understanding of flood risk in complex social and ecological settings across many spatial and temporal scales (Government of India, 2008: 31-32). This chapter focuses more on the non-structural measure of disaster preparedness, which is discussed in forthcoming paragraphs.

Disaster preparedness identifies the steps necessary to increase the probability of preventing or buffering the hazardous effect associated with a disaster event. Preparedness strategies are developed through hazard identification, mapping, analysis and risk valuation (Ejeta *et al.* 2015). Effective preparedness therefore results in reduced vulnerability, increased mitigation levels, enabling timely and effective response to a disaster event and minimising the reclamation period from a disaster and intensifications community resilience (IFRCS, 2000). Disaster preparedness involves predicting and taking preventative measures prior to a looming threat, organising the delivery of timely and effective rescue, relief and assistance, securing resources by stockpiling supplies and earmarking funds (Kotze and Halloway, 1996). It also involves the systematic testing of warning systems and plans for evacuation, to minimise potential loss of life and physical damage, the education and training of officials (intervention teams) and the population at risk and the establishment of policies, standards, organisational arrangements and operational plans to be applied following a disaster (Kent, 1994:11).

Kent (1994: 12) argues that disaster preparedness is an active, on-going process, since the preparedness plans are dynamic ventures which need to be reviewed, modified, updated and tested on a regular basis (see Table 4.1). This therefore means that the plans should not be shelved to be used in the field when a disaster strikes, because they may be useless in the field as things constantly change on the ground.

Table 4.1: Disaster Preparedness Framework (Kent, 1994: 16)

Vulnerability Assessment	Planning	Institutional Framework
Information Systems	Resource Base	Warning Systems
Response Mechanisms	Public Education and Training	Rehearsals

The first stage of the framework is *vulnerability assessment*, which is essential to creating a disaster management plan and entails identifying areas that are at risk of a hazard, such as drought or flood prone areas. Kent (1994: 16) argues that it is a continuing, dynamic process of people and organisations assessing the hazards and risks they face and determining what they wish to do about them, using structured data collection to understand the levels of potential threats, needs and immediately available resources. The information collected relates to infrastructure in terms of the extent of expansion, vulnerability faced by the area and a "map" of available structures (roads), which may be valuable in times of emergencies (IFRCS, 2000). The second set of information is socioeconomic data, which shows the causes and levels of vulnerability, demographic shifts and types of economic activity. Therefore, this stage is focused on the likely effects of potential hazards, relief needs and available resources.

Planning is the second stage of disaster preparedness and entails working out agreements between people or agencies as to who will provide services during the emergency to ensure an operational, harmonised response. It therefore stimulates an on-going interaction between parties which may result in written, usable agreements. According to IFRC (2000), the planning process entails the involvement of the planners as the experts and the participation of local people and grass-roots organisations, including NGOs, to generate commitment from the various stakeholders. The plan should not be *ultra vires* to the legislation of the nation and should ensure that funds are available for its implementation.

An *institutional framework* seeks to provide a coordinated structure for the carrying out of disaster risk management activities. The institutions involved should have good working relationships and their roles and responsibilities should be clearly defined and each institution should have the expertise to execute them (APFM, 2011: 13-14). *Information systems* entail setting up early warning systems which should be linked with grassroots information so that it becomes relevant to and is not be ignored by the local community within which it will be implemented. The UNDP (1992:19) argues that disasters occur but people become vulnerable to them simply because they do not know how to get out of harm's way or take protective measures. Kent (1994: 27) argues that information exchange systems between institutions should be effective and well understood by all concerned parties.

A resilient *resource base* needs to be created which identifies resources for disaster relief and recovery and includes setting up a fund to purchase resources that cannot be easily stockpiled such as medicine, insurance and facilitating the coordination of bilateral donors and stockpiling relief materials such as food reserves (Kent, 1994: 28). *Warning systems* inform authorities of the impending danger of floods. APFM (2011: 9) states that longer lead time will provide sufficient time to consider and effect a number of responses and the warnings must be unambiguous, easily understandable and in the local language and channelled through a legally designated authority. The warning systems should be affected at intervals to ensure that the disaster preparedness plan is implemented about 72-, 36- and 12-hours intervals before the disaster strikes, so that many people can hear it as soon as possible.

Response mechanisms have to be defined by the plan in terms of the institutions that are responsible for executing them. These include: evacuation and search and rescue procedures, preparation of emergency reception shelters and activation of distribution systems (UNDP 1992). *Public education and training* are concerned with providing people with the information needed to mitigate the effects of disasters. Public education on disasters, workshops, outreach programmes and radio and television can be used to educate the public, showing what they can do to help themselves when disaster strikes (IFRC, 2000). Ward (1991: 23), states that it is important that relief workers and those within the institution be trained to help others and community leaders be shown how to prepare their communities for the disasters. APFM (2011: 7) argues that outreach efforts should be made to minorities and ethnic groups, as their mobility is limited due to cultural, social or economic constraints, especially women and children, as they are extremely affected by natural disasters.

Rehearsals/emergency drills entail separate training programmes from central to local authorities to the local community that test the system as a whole to expose gaps that may have been overlooked (Kent, 1994: 34; see Box 4.1). The National Fire Protection Association (NFPA, 2000) states that this helps to ensure an orderly evacuation rather than a speed evacuation that may result in more harm than the disaster itself.

Box 4.1: Rehearsals (Kent, 1994: 35)

A two-day exercise was held in November 1982 in Yugoslavia which simulated an emergency at the Krsko nuclear power plant. More than 70,000 people took part, including 8,000 officials in on-site and off-site reaction groups and organisations. As part of the exercise, a village close to the plant was selected for the rehearsals to demonstrate complete evacuation, while people in the rest of the area were taught how to find safe shelter around their areas. Defences were made to avert contamination of food supplies and fire-fighting demos were done under full radiological pollution control while traffic circulation controls and decontamination amenities were created.

The Case of Beitbridge

Beitbridge Town was a creation of S.I. 120 of 2007, Proclamation 3 of 2007. According to the Metrological Office of Zimbabwe, it is a low lying floodplain area that is prone to and most affected by floods within Zimbabwe (See Figure 2 which is the map showing the cardinal points of its boundary being, A- 922392 (middle of Limpopo River), B - 931428 (close to Beitbridge Station), C - 961431 (on the railway line), E - 970485 (on the Beitbridge-Masvingo Road), F(940470 (along the Masvingo-Beitbridge Road), G - 083478 (on the Bulawayo-Beitbridge Road) and H- 017436 (middle of the Limpopo River).

Figure 4.2: Map of the Beitbridge Town (Fieldwork, 2016)

The town was established as a port of entry/exit into/from South Africa. It received the town status in 2007, prior to which it acted as the administrative centre for the whole of Beitbridge District under the Beitbridge Rural District Council. Consequently, it became the main urban centre in the Southern part of Matabeleland South, attracting rural folks who migrated to escape perennial droughts in the surrounding communal areas, contributing to its growth in size. According to ZimStat (2012), the population of Beitbridge is 42 137 with 11 959 households and a transit population of 10 000 people.

Further rapid growth of the town is attributed to it being a Border post which is a regional port of entry and exit. According to *Herald* (12/03/15) it is one of the busiest inland border posts in Southern Africa, as it handles most of the goods that are in transit to central and northern parts of Africa. This is because it facilitates trade within the Southern African Development Community (SADC), Save the Children, 2009). According to *Herald* (12/03/15), a total of about 170 000 people, 2 100 buses, 25 000 private cars and 15 000 trucks pass through the town every month. This resulted in the construction of a new larger border post being constructed in 1994, followed by subsequent 4 extensions to the border, over the years and the construction of government offices and houses and flats to accommodate civil servants (*Herald*, 12/03/15). Therefore, the population growth rate of Beitbridge exceeds the national annual rate by 5.72 % due to in-migration and consequently the population is likely to double in the next 10 years. This therefore calls for effective planning and preparedness for disasters to protect the generations to come.

Methodology

This chapter is based on research done using literature and document review, which interrogated existing published and unpublished documents and databases such as newspapers and journals. Document review was particularly useful in providing estimates of relevant parameters such as current data about what is happening on the ground in terms of floods in Beitbridge and past trends of the floods. However, important to note is that data on floods in Beitbridge is limited since not much research has been done or documented in this regard. The other form of data collection was through informal interviews with key informants, who possess first-hand knowledge about the community, to get a greater understanding of the issues at hand.

Results and Discussion

There are two types of floods that affect the Beitbridge area, the first and most frequent type of floods are the seasonal floods, usually in January or February every year. The second type is the cyclone-induced floods, which are increasing in frequency, as seen in the occurrence of cyclone Eline in February 2000 and of cyclone

Japhet in March 2003 (GMO/GWP, 2004). In 2013, torrential rains hit Beitbridge, affecting Beitbridge Town and rural areas such as Tshasvingo, Tshitulipasi, Mawale and Tshikwalakwala (*Newsday*, 26/01/2013), leaving many people dead and injured. The floods also resulted in the deaths of 3 children, after the houses they were sleeping in collapsed and 2 people drowning in Bubi and Tshitulipasi Villages. Nehanda Radio (22/01/2012) observes that the floods were so severe that the Airforce had to be dispatched to help to assist stranded villagers and in Beitbridge Town, the border post was closed between 12 midnight and 3 am after the Limpopo River had flooded. The floods also damaged the roads, sweeping bridges away and forcing the closure of the district hospital.

Beitbridge Town was also hit by a massive flash flood on the 11 March 2016, affecting approximately 1000 people and about 238 households. 300 of these people in these households were women, some of them disabled, pregnant and others lactating mothers and about 285 were children (UNFRA, 2016). The Relief Web (2016) argues that the localized floods were caused by torrential rains associated with El Nino phenomenon and the main area affected by the floods was Dulibadzimu suburb, which is one of the border towns of Beitbridge. The area received around 107 mm of rainfall, which is the highest recorded per day in the last 7 years, which led to the majority of the residence to take refugee on rooftops as the rainfall reached waist level (Herald, 16/03/16). Affected were 76 houses located near the bus terminus and, as well some 225 houses in Dulibadzimu and the council hostels. The aid received by the victims include: clothes, food items, books, blankets and 660kgs of rice from the Minister of State for National Security Kembo Muhadi, Lobels, churches, clearing companies and non-governmental organisations such UNFPA and the Red Cross. According to *Herald* (16/03/16) the effects of the flood include damaged houses, general hospital, roads, stadium, social amenities, and electricity lines and burst sewer pipes resulting in the water supply lines being cut short. Some residents of the town lost their life savings, including one who lost $18 000 and saw his cars been washed away by the flood. Other valuables lost included televisions, fridges, beds and sofas (*The Mirror*, 20/03/16).

The public affected by floods in the town tend to be angered and feel wronged by their city fathers, since the Government is perceived to pay a lot of lip service after disasters occur without taking corrective action afterwards. According to *The Mirror*

(20/03/16), many residents blame the flood disaster on local leadership, the town council and local civil and political leadership, which focuses more on politics and elections rather than serving the people or their development needs. They argue that the flood issue is a time bomb that could have been solved a long time back before it posed a serious damage to the people. The residents of the town argue that the town is important to the country as it brings millions of dollars every year through taxes and import duties and the least that the government can do, therefore, is to pour back some of the money made to develop the area, which is characterised by dilapidated structures, poor water and sewer reticulation and poor drainage systems (the Mirror, 20/03/16).

In terms of policy and legislative framework, the disaster preparedness of Zimbabwe is found lacking. Although ZIMASSET (2013-2018) upholds the need to establish a national climate change policy to help deal with climate-related disasters, little seems to be done on the ground. According to *Herald* (01/04/16), the National Climate Change Response Strategy was crafted in 2014. Two years on, it has not been completed. A Disaster risk expert commented on *Herald* (01/04/16) that "unless there is a comprehensive climate change policy and strategy that is directed at national level through the relevant ministries, disaster preparedness will remain compromised".

As regards the CPU procedures, a key informant working in the CPU indicated that the department is required to produce operational emergency preparedness plans initiated during disasters and should be reviewed regularly at least once a year. The plans indicate the alert mechanisms, evacuation procedures, stock of resources available (material and human) and the contact details of manpower. The District Civil Protection Coordination Committee (DCPC) in Beitbridge is grouped into sub-committees namely,

- Food Supplies and Food Security: This provides food for the victims and is there to ensure that food supplies are stored and ready for use before the floods hit. This is facilitated by Grain Marketing Board (GMW).
- Health, Nutrition and Welfare: The CPU works hand in hand with the Health services within the district to attend to the injured, while the Social Welfare looks at the needs of flood victims, providing social care to the wounded throughout and after the crisis.

- Search, Rescue and Security: This is chaired by the Zimbabwe Republic Police and the Defence Forces, who are responsible for searching, rescuing and relocating flood victims and providing security during flood crisis.

- International Cooperation and Assistance: the CPU asks for international assistance from the United Nations and other NGOs when the disaster they are facing exceeds their capacity to respond to it.

The other important government bodies that the CPU works with are the Zimbabwe National Water Authority (ZINWA) and the Meteorological Department, which form the early warning system. They are responsible for predicting the weather and, possibly, flood events so that potential victims are evacuated before or during the flood event and the public are kept updated as events unfold. ZINWA's other role is to determine the current water situation in terms of its quantity and quality to meet human needs and also to look at the water infrastructure and carry out the necessary repairs to ensure that water is made available to the community, even drilling boreholes to meet the demand.

The level of preparedness of the Beitbridge area is low, as its early warning is weak in turn. According to Madamombe (2007), the early warning system of the Metropolitan Department and ZINWA have two problems: the first one is that the lead-time between the flood forecast and the flood event used for meteorological forecasts provides very short forecasts accurately and this may not allow the CPU enough time to take appropriate action such as early evacuations. The second one is that there is low level of accuracy when it comes to the forecasts. Owing to previous false alarms from the Met Office, people are no longer taking its forecasts seriously as demonstrated during the Eline cyclone, when many ignored its reports (GMO/GWP, 2004: 5). Worse still, there is little information about the disasters on the ground, making proactive planning impossible. According to a disaster risk expert in *Herald* (01/04/16), the current state of Zimbabwe's environmental information management systems is such that there is lack of continuous local environmental monitoring and local environmental data collection and analysis, making it very difficult to make acceptably accurate predictions and model the environment".

Another challenge faced by CPU is the general collapse of the institution due to economic hardships making it incapacitated to ensure adequate disaster preparedness. According to a key informant from the Department, there is inadequate sharing of information and too much bureaucratic red tape, as shown by the chat below. This has affected the ability of the department to develop a database on disaster risk reduction. Information is exchanged through the sharing of reports, minutes and newsletters between the CPU, local authority, community, the meteorology department and other government institutions. There is therefore no efficient modern infrastructure to facilitate communication, networking and interconnection amongst stakeholders, making it difficult to create effective disaster preparedness plans, as availability of information is limited and that received may in most cases be outdated. Figure 4.3 shows the flow of information between institutions for flood management emergencies in Zimbabwe.

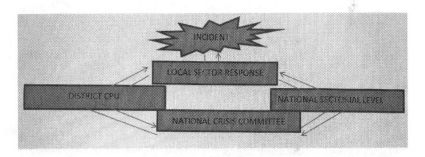

Figure 4.3: The Flow of Information (Operational Manual Management of Flood Emergencies in Zimbabwe and APFM, 2004: 6)

According to the informant at CPU, the unit is proactively engaged in creating public awareness campaigns about flooding, drowning and lightning hazards, which are carried out towards and during the wet season. In April 2014, the Civil Protection Unit launched a community-based disaster risk management programme in low lying areas in Beitbridge, which are constantly affected by the flood. The CPU District Chairman argues that the programme's main objective was to empower the community with disaster management and preparedness knowledge in terms of disaster reduction, emergency preparedness and response planning and recovery at ward level to save lives and property. The campaigns are carried out through the press, pamphlets and road shows, but due to financial constraints, they cannot make use of the electronic media, which is a more far reaching method than the ones used.

Although public awareness is an important part of disaster preparedness, it is not the only aspect of the overall framework. This is especially true of its response mechanism, which has serious shortcomings. This was evident when the floods hit the area in January 2013, forcing most inhabitants of the town to flee where the floods had not caused much damage, after the authorities had turned a blind eye to their plight (*Newsday*, 26/01/13). Nehanda Radio (22/01/13) also indicates that many of the victims were stranded and others were marooned within the park, which is management by the Department of Parks and Wildlife and had to be rescued the day after the floods hit. This shows serious shortcomings on the part of the authorities.

Although it is encouraging that the National Civil Protection Fund was established to finance the development and promotion of civil protection measures, the effectiveness of such a provision is dependent on the Government of Zimbabwe's fiscal budget (Save the Children, 2009). This is one of the challenges that have been cited by the CPU: it has not been able to effectively provide protection to the public owing to insufficient financial commitment on the part of the government. The department has therefore been unable to secure funds to install the appropriate infrastructure and well advanced and planned educational programmes to the public. The lack of timely provision of funds has affected the Department and the local authority, reducing their capacity to be proactive before the disaster strikes. According to *Herald* (12/01/16), the CPU received just $300 000 for the entire year from the National Budget, the bulk of which went towards current expenditure such as salaries and administrative costs, which were given priority over disaster planning. The CPU Director in *Herald* (12/01/16) was quoted as stating that in 2013, the Unit did not financial resources and could not even afford to buy fire-fighting equipment and its capacity to respond in the aftermath of disasters was undermined.

In the area under study, it can be noted that after the floods had hit the area in March 2016 that the Town Secretary of Beitbridge stated that they have now set aside $73 000 to address the problems of drainage and another $40 000 for construction along the Wamalala River, which passes through the area between the stadium and the district hospital. Given that the drainage problem is well-known in the area, this raises the question of why the authorities act after the flood and not before. According to *The Sunday Mail* (27/03/16), Beitbridge Town has the worst drainage systems in the country, a phenomenon which could have been dealt with before the disaster occurred. Even one of the residents of the Town was quoted in *The Mirror* (20/03/16) as saying "the town's drainage system has been questionable for a long time, but the Government has not done anything about it; it's stupid because we are talking about human life". Disaster preparedness also entails constant monitoring of the plan in terms of availability of transport fleets and warehousing facilities to ensure that the implementation of the plan is efficient and enable timely relief to be provided (Kent 1994: 26). This is another area in which the planning authorities in Beitbridge fall short. It was during the flood event of March 13 that the Town Council Chairman stated that they were now engaging local businesses with warehouses to accommodate the household goods of victims. This therefore shows that no preliminary studies were done to assess the resource base of the area and make arrangements with the businesses in time.

The CPU is found wanting in terms of stockpiling of relief materials. According to the informant there, the GMB is mandated to stockpile food as food security reserve, but its effectiveness has been reduced and to be able to stockpile owing to the economic instability the nation has been facing since 2009. This can be noted in the revelation by *The Sunday Mail* (27/03/16) that vulnerable families in Beitbridge are being forced to contribute money to buy maize from the GMB, contrary to the government directive that they should receive the food aid for free.

The area is guided by the Local Development Plan of 1984, which was amended in 1992 to provide for a new border post and ancillary facilities. In 2003, the town prepared a Local Development Plan which was not approved, because the consultation process was biased and did not fully represent the values and needs of all stakeholders and did not come up with environmental proposals in line with the Environmental Management Act. The Zimbabwe Institute of Regional and Urban Planning (ZIRUP) President is quoted in *The Sunday Mail* (27/03/16) stated that most small towns like Beitbridge are ill equipped to deal with disasters as they rely on old plans that have not been reviewed in a long time, operating without sufficient expertise in town planning issues. This therefore raises the question of whether there is appropriate environmental planning in the area in the absence of a proper planning framework in the form of up-to-date Local Development Plan. The one currently operating has outlived its lifespan and is no longer relevant to tackle the ever-changing environment the town is facing in terms of rapid urbanisation and climate change. This, in turn, means that the Disaster preparedness Plans created by the CPU are not aligned with the Local Development Plan, which should inform the development and establishment of settlements within the area. This could explain why there is no proper planning for those who are homeless and live in the streets, which is one of the tasks of urban planners, who are supposed to create inclusive urban areas. According to *Herald* (16/03/16), one of the groups affected by the floods is the vagrants and beggars who live along a stream in Dulibadzimu.

The disjuncture of local planning in the area and disaster preparedness is evident in the lack of alignment between the two. According to *The Sunday Mail* (27/03/16), Beitbridge Town is affected by poor town planning, as the worst affected houses are built on the course of the Wamulala River, a tributary of the Limpopo River, such that heavy rains upstream will cause flooding in the Town. Despite the fact that the area is known to be a flood zone, the town council continues to allocate stands right in the middle of the river course (*The Sunday Mail*, 27/03/16), showing that disaster preparedness in the area does not inform local planning in the area.

Another example of lack of environmental planning in Beitbridge is the poor waste disposal system. Because the area is operating under the Local Development Plan of 1984, much of the current developments that have occurred in the area lie outside the catchment areas of the existing sewage ponds provided under the Plan, rains result in over-spilling of the ponds into the Limpopo River, itself an environmental hazard. According to Relief Web (21/09/15), this has resulted in cholera epidemic which, in 2008 and 2009, affected more than 25 000 and it took 2 years to recover from it. Save the Children (2009) argues that the town is filled with filth and has limited basic services, like water, health, telephone system and electrical power and the people can go for several months without water due to a malfunctioning and under-capacitated water treatment plant. Therefore, without proper planning for disposal of waste and sewer lines, the epidemic accelerates as floods destroy the pipes and worsen the situation. The floods in March 2016 did exactly that. *Herald* (16/03/16) reported that sewer lines burst in Dulibadzimu. This forced the authorities to close water supplies. This means that more sewage is released to the River resulting in water pollution, affecting areas like Mozambique, which is downstream. According to Radio Dialogue (17/03/16), the council embarked on a clean-up exercise after the floods to prevent the outbreak of diseases like malaria.

Conclusion and Recommendations

This study sought to explore the gaps and constraints in environmental planning and disaster preparedness of Beitbridge, with a view to enhancing the capacity to of local authorities to effectively respond to disasters. The study showed that the early warning systems and response mechanisms relating to floods in the area are weak and that there is little or no proactive planning. The research therefore recommends the improvement of communication so that all potential victims can be reached. There is need to establish an effective coordinated flow of information between institutions and for constant monitoring and evaluation of the disaster preparedness plan, especially in terms of vulnerability assessment, to increase the responsiveness of the local authorities to the disasters. Specifically, there is need to:

- improve on communication so that all potential victims can be reached, especially with regard to how people respond to floods;
- establish an effective coordinated flow of information between institutions;
- improve the existing capacity to do forecasts so that they become more accurate;
- be proactive in terms of planning and creating working agreements between agents to enable operational and timely response to disasters;
- strengthen the response mechanisms of the CPU and the local authorities;
- engage in rehearsals to give both the public and the authorities hands on experience on how to act during a disaster and to strengthen the whole system of disaster preparedness within the local area;
- constantly monitor and evaluate disaster preparedness plan especially in terms of vulnerability assessment;
- environmental policies to be established for the protection of water courses and disposal of waste;

77

- align the Local Development Plan with the Disaster preparedness Plan for the Beitbridge Town area.

References

Associated Programme on Flood Management (APFM). (2011). *Flood Emergency Planning.* Integrated Flood Management Tools Series No.11, World Meteorological Organisation

Chambers, R., and Conway, G. (1998). *Sustainable Rural Livelihoods: Practical Concepts for the 21ˢᵗ Century.* Institute of Development Studies. Sussex: United Kingdom.

Kadirire, H. (April 16, 2015). Sort Out this Beitbridge Mess. *Daily News*, Zimbabwe. Available online: https://www.dailynews.co.zw/articles/2015/04/16/sort-out-this-beitbridge-mess. [Accessed on 10 July 2018]

Department of Civil Protection, Zimbabwe Hazard Profile. Civil Protection Department Available online: https://www.researchgate.net/publication/282488178_Disaster _legislation_a_critical_review_of_the_Civil_Protection_Act_of _Zimbabwe, [Accessed on 10 July 2018]

Ejeta, L. T., Ardalan, A., and Paton, D. (2015). Application of behavioural theories to disaster and emergency health preparedness: A systematic review. *PLoS currents*, 7.

Ferris, E., And Petz, D. (2013). *In the Neighbourhood: The Growing Role of Regional Organisations in Disaster Risk Management.* Brookings.

Government of India. (2008*). Natural Disaster Management Guidelines: Management of Floods.* Rashtrapati Bhawan, New Delhi, Government of India Printers.

Government of Zimbabwe. (1996). Regional Town and Country Planning Act (Chapter 29: 12), Government of Zimbabwe: Harare

Government of Zimbabwe. (1998). Water Act 1998, Government of Zimbabwe, Harare.

Government of Zimbabwe. (2001). Civil Protection Act, Chapter 10:06. Government of Zimbabwe, Harare.

Government of Zimbabwe. (2003). Meteorological Bill 2003, Government of Zimbabwe, Harare.

Government of Zimbabwe. (2013). Constitution of Zimbabwe Amendment (No 20) Act 2013. Fidelity Printers and Refiners, Harare

Government of Zimbabwe. (December, 2012). Zimbabwe National Contingency Plan (December 2012 – November 2013). Government of Zimbabwe and United Nations, Harare.

Gwimbi, P. (2009). Linking rural community livelihoods to resilience building in flood risk reduction in Zimbabwe. *Jàmbá: Journal of Disaster Risk Studies*, 2(1), 71-79.

Herald. (April 01, 2016). Formulate Disaster Management Policy. Zimpapers, Harare

Herald. (March 12, 2015). Beitbridge on the Mend. Zimpapers, Harare

Herald. (March 13, 2016). Flood Hit Beitbridge. Zimpapers, Harare

Herald. (March 16, 2016). Beitbridge Flood Victims Rise to 780. Zimpapers, Harare

Herald. (March 18, 2016). Police Warn of Flash Floods. Zimpapers, Harare

ICLEI. (2015). Resilient Cities Report; Global developments in urban adaptation and resilience. Bonn, Germany

International Federation of Red Cross and Red Crescent Movement (IFRC). (2000). Disaster Preparedness Training Programme. Introduction to Disaster preparedness. IFRC.

Kent, R. (1994). *Disaster Preparedness: Disaster* Management Training Programme. United Nations Development Programme, Wisconsin

Knutsson, P. (2006). The sustainable livelihoods approach: A framework for knowledge integration assessment. *Human Ecology Review*, 13(1), 90-99.

Leichenco, R. (2011). Current Option in Environmental Sustainability. *Elsevier.* 2(3), 164 – 168

Madamombe, K. (2007). Zimbabwe: Flood Management Practices: Selected Flood Prone Areas Zambezi Basin. WMO/GWP Associated Programme on Flood Management. Zimbabwe National Water Authority, Research and Data Department, Harare

Nehanda Radio. (January 22 2013). Floods Close Beitbridge Border Post. Available online: http://nehandaradio.com. [Accessed on 12 February 2018]

Newsday. (January 26, 2013). Beitbridge Flood Victims Stranded. *Newsday*, Zimbabwe

Radio Dialogue. (March 17, 2016). Clean Up for Flood Hit Beitbridge. Radio Dialogue, Zimbabwe

Relief Web. (March 24 2016). UNFPA Restores Women's Dignity in Flood Affected Beitbridge. UN. Available online: http://reliefweb.int/disasters. [Accessed on 17 June 2018]

Relief Web. (September 21 2015). Riddinf Beitbridge of Cholera: Zimbabwe's Hardest Hit District Fights Back. UN. Available online: http://reliefweb.int/disaster. [Accessed on 17 June 2018]

SADC. (2011). Southern Africa Regional Flood Update: 20 January 2011. SADC, Botswana

SADC. (2015). SADC at 35; Success Stories, Volume 1. SADC, Botswana

Save the Children Alliance. (2009). *Rapid Assessment of Protection Issues within Zimbabwe's Cholera Epidemic and Response.* Save the Children Alliance, UK

Scoones, I. (2000). Sustainable Rural Livelihoods: A Framework for Analysis. IDS Working Paper No.72. University of Sussex: United Kingdom.

Spaliviero, M., Dapper, M. D., M Mannaerts, C., and Yachan, A. (2011). Participatory approach for integrated basin planning with focus on disaster risk reduction: the case of the Limpopo river. *Water*, 3(3), 737-763.

Sunday Mail. (March 27, 2016). No bridge over Troubled Waters... as GMW defies Government Grain Directive. Harare, Zimpapers.

The Mirror. (March 20, 2016). Disaster as Flash floods destroy Beitbridge. *Mirror*, Masvingo, Zimbabwe

Tyler, S., & Moench, M. (2012). A framework for urban climate resilience. *Climate and development*, 4(4), 311-326.

United Nations Development Programme (UNDP). (1992). *An overview of disaster management, 2nd, Disaster Management Training Programme*, UNDP/UNDRO, New York

United Nations Habitant (UNHABITAT) (2012). UN Systems Task Team on the Post 2015 UN Development Agenda: Sustainable Urbanisation. United Nations, Geneva

United Nations Office for Disaster Risk Reduction (UNISDR). (2012). *How to Make Cities More Resilient: A Handbook for Local Government Leaders.* United Nations, Geneva

United Nations Office for the Coordination of Humanitarian Affairs (OCHA) (2014). Flood Situation Report No 1-6 2014. Harare, UN OCHA, Mt Pleasant.

World Meteorological Organisation., and Global Water Partnership (WMO/GWP). (2004). *Zimbabwe; Flood Management Practices-Selected Flood Prone Areas Zambezi Basin*. WMO/GWP

Zimstat. (2012). Census 2012 National Report. Harare, Government of Zimbabwe.

Chapter 5

Inside the Climate Change Policy-Making in Zimbabwe

Kenneth Odero

Introduction

The presence (or absence) of national climate change policy frameworks in a given country is often taken as a testament of its capacity to formulate climate change policies and response strategies. Using this logic, Zimbabwe's slow pace in formulating a national climate change policy and response strategy has been construed as symptomatic of a 'lack of capacity'.[1] Such a binary view of climate policy-making is flawed for the following reasons. First, it fails to appreciate national policy-making interests and priorities. Second, it is ahistorical and totally neglects embedded policy-making capabilities that come with experience. Given Zimbabwe's deep policy-making know how and experience, explanation for its relatively slow pace in national climate change policy-making must lie elsewhere. This chapter points to Zimbabwe's constricted policy space, disengagement with the international community and the contradictions around the climate change agenda as the real reasons for the 'slow' pace in formulation of a national climate change policy and response strategy.

Zimbabwe's National Climate Change Policy Regime

The Government of Zimbabwe, through the Ministry of Environment, Water and Climate put out a call for expression of

---

[1] This perspective has morphed to an extent that capacity building has become one of the single most important identifier of climate change action/inaction. For a perspective on this phenomenon, see a recent review of intended nationally determined contributions (INDCs) submissions under the UNFCCC (Umemiya, Goodwin and Gillenwater, "How intended did nationally determined contributions (INDCs) address capacity building? Implications for future communication by Parties on capacity building under the Paris Agreement", *IGES Working Paper*, July 2016).

interest for development of climate policy for Zimbabwe in February 2015. The call stated that the "lead consultant shall facilitate, coordinate and harmonise the process of the development of the National Climate Policy". The Government of Rwanda commissioned the development of a National Climate Change and Low Carbon Development Strategy for Rwanda in 2009. Box 5.1 is a matrix of a sample of some countries with national climate change policies. The interpretation is stark: Zimbabwe lags behind most other countries in developing a national climate change policy and response framework.

Box 5.1: Examples of National Climate Change Policies (Author's own compilation)

- Brazil's National Plan on Climate Change (Government of Brazil 2008)
- Climate Change National Adaptation Programme of Action of Ethiopia (The Federal Democratic Republic of Ethiopia 2007)
- China's National Climate Change Program (Government of China 2007)
- Indian National Action Plan on Climate Change (Government of India 2008)
- Mexico National Strategy on Climate Change (Estados Unidos Mexicanos 2007)
- South Africa's National Climate Change Response Strategy (Government of South Africa 2004)
- Tanzania's National Adaptation Program of Action (United Republic of Tanzania 2007)
- Uganda National Climate Change Policy (The Republic of Uganda 2012)
- National Policy on Climate Change for Namibia (Republic of Namibia 2010)
- Malawi National Climate Change Policy (Government of Malawi 2013)

In a presentation at the First Climate Science Symposium of Zimbabwe held in 19-21 June 2013 in Harare, I pointed to the strong, credible and irrefutable evidence that the effects of climate change were already having in parts of Zimbabwe. I also cautioned that unless urgent action was taken, climate change is likely to have significant ramifications for Zimbabwe's economy in the future. According to the African Union *Guidebook on Addressing Climate Change Challenges in Africa* (AMCEN, 2011), national policies frameworks are useful instruments for climate change adaptation and mitigations.

Despite lagging behind most other nations in terms of developing a national climate policy and response strategy, Zimbabwe has been engaged in the global frameworks for climate change. According to the UNFCCC (2011), status of submissions, Zimbabwe submitted the Initial National Communication (INC) 25 May 1998 and the Second National Communication (SNC) on 31 January 2013. According to *Zimbabwe Second National Communication to the United Nations Framework Convention to Climate Change* (Republic of Zimbabwe, 2013), the country was among the first to sing and ratify the UNFCCC at the United Nations Conference on Environment and Development held in Rio de Janeiro in June 1992. Zimbabwe also acceded to the Kyoto Protocol on 28 September 2009. All these point to Zimbabwe's engagement with climate international policy.

The SNC goes on to highlight some of the initiatives undertaken since it ratified the Convention. These include Technology Transfer Needs Assessment; Global Environment (GEF) Fund/Small Grants Programme; National Capacity Needs Self-Assessment for implementation of the Multilateral Environment Agreements; United National Development Program/GEF Medium Size Project on Coping with Drought and Climate Change; Climate Change Awareness Workshops; among others. Nonetheless, there is not a single reference to national climate change policy frameworks. This is not surprising given the constricted policy space occasioned by the Zimbabwean crisis, as well as the country's broader disengagement with the international community on a number of issues, most notably land reform, human rights and electoral politics.

Revisiting Global Climate Policy (In) Action

In a recent report, the Intergovernmental Panel on Climate Change (IPCC) warns climate change will create new poor between now and 2100, in developing and developed countries and jeopardize sustainable development' (IPCC, 2014).[2] Lack of decisive action to save the world against climate change has never been due to the lack of evidence or scientific basis of global warming. The main obstacle to corrective action has been vested interests. One of

---

[2] "Climate Change 2014: Impacts, Adaptation and Vulnerability": Chapter 13, http://ar5syr.ipcc.ch/resources/htmlpdf/ipcc_wg3_ar5_Chapter4/, p. 796

the clearest examples of this was the disdainful manner with which the *Limits to Growth* (Meadows *et al.* 1972) report was received. Lack of serious action to move away from the business-as-usual path of population, industrialisation, pollution, food production and resource depletion marked a historical turning point in the global warming crisis. The collective failure to pay heed to scientific evidence on climate change is explained by vested interests of large powerful corporations, more than anything else.

The unbridled pursuit of materialism, imperialism, empire building and accumulation on a world scale characterized by global corporatism has historically been responsible for pushing the planet dangerously close to its tipping point. The world is surviving on borrowed time, or so it seems if one takes predictions by the climate scientific community, most notably the IPCC, seriously. But, in spite all available evidence, the climate change narrative continues to be manipulated to suit narrow economic and political interests. For example, climate change deniers, backed and financed by powerful corporations have taken the slightest opportunity to shoot down, often without hard empirical evidence, scientifically grounded arguments on global warming and its effects on climate. This doubling down on anthropogenic climate change by global corporate interest is a recurring theme of critical discourse in the global South in general and more particularly within the commanding heights of Zimbabwe's policy-making community.

Enter the Land Reform Programme

Ideologically, the Zimbabwe African National Union Patriotic Front (ZANU-PF) – the ruling party – is left of centre. Despite is socialist leanings, ZANU-PF did not have a serious rift with the West during much of the first two decades of post-independence (i.e., 1980-1997). For example, Zimbabwe, like most sub-Saharan African countries adopted the neoliberal Economic Structural Adjustment Policies (ESAPs) prescribed by the Bretton Woods at the start of the 1990s. However, this partnership was severely tested following the disastrous effects of ESAP. As I have argued previously, the adoption of ESAP in 1991 was a major turning point in the development of Zimbabwe.[3] More ominous cracks

---

[3] In an effort to raise the rate of economic growth, the government turned to external borrowing to finance investment, reduced bureaucratic requirements for

surfaced following Zimbabwe's land invasions in the early 2000s and subsequent implementation of the Land Reform and Resettlement Programme introduced to redress 'historical land injustice'[4].

The Land Reform and Resettlement Programme was used by western countries, led by the United Kingdom (UK) government under Prime Minister Tony Blair, to impose sanctions on Zimbabwe to try and force political reforms in the country. The leadership in Zimbabwe read this as mischief designed to effect regime change and as they dug-in and hardened their position, the economy took a knock that led to a massive contraction. With the economy of a free fall, the government switched its attention to dealing with crisis at hand. As a result, climate change became a peripheral concern and less of lesser priority to the government. This constricted policy space is one of the main reasons that contributed to the relatively slow pace in formulating a national climate change policy and response strategy for Zimbabwe. But another, perhaps equally important complicating factor is the intractable international climate policy negotiations.

The Conference of Parties (COP) 'merry-go-round'

Another equally important factor influencing climate change policy-making in Zimbabwe was the contradictions around the climate change agenda. Although Zimbabwe was among the first counties that signed the Kyoto Protocol following its adoption at the Third World Climate Conference (COP3) in 1997, protracted delay in reaching an international climate agreement on the one hand and the tense relationship between Zimbabwe and the west reinforced the latter's lacklustre approach towards responding the climate change agenda. It is useful to recall that the Kyoto Protocol came into effect in 2005. By then, the Zimbabwean economy was

---

investments and the conduct of business and liberalised trade in order to improve the availability, price and quality of goods in the local market. This policy shift, particularly the half-hearted manner in which the government implemented it, was largely responsible for the economic upheaval that followed. See Odero, K. (1999) "Local Authorities' Response to Restructuring". Department of Rural and Urban Planning *Working Paper*, University of Zimbabwe

[4] For more on Zimbabwe's land reform programme, see S Marimira and K Odero (2003). *An analysis of institutional and organisational issues on fast track resettlement: the case of Goromonzi District.*

showing serious signs of weakening and the relationship between Harare and the West was anything but cosy. To achieve the objectives of the Kyoto Protocol, States negotiated the establishment of a carbon market with the trading of greenhouse gases (GHGs). The notion of a carbon market is one area that is considered controversial and which has shaped Zimbabwe's (as well as other countries') disposition to climate policy. I return to this point later.

One of the key principles undergirding the Kyoto Protocol was 'differentiated responsibilities but a common goal'. A legal brief of the Centre for International Sustainable Development Law (CISDL) based at in McGill University explains this principle as having evolved from the notion of the 'common heritage of mankind' and is a manifestation of general principles of equity in international law. The principle recognises historical differences in the contributions of developed and developing States to global environmental problems and differences in their respective economic and technical capacity to tackle these problems. Despite their common responsibilities, important differences exist between the stated responsibilities of developed and developing countries. The principle of common but differentiated responsibility includes two fundamental elements. The first concerns the common responsibility of States for the protection of the environment, or parts of it, at the national, regional and global levels. The second concerns the need to take into account the different circumstances, particularly each State's contribution to the evolution of a particular problem and its ability to prevent, reduce and control the threat (CISDL, 2012).

However, at COP7 held in Marrakech, Morocco in 2001, certain developed countries sought equally responsibility in terms of emissions quotas. Developing countries represented by the Group of Seventy-seven (G77), of which Zimbabwe is a prominent member, considered this request inappropriate coming from countries known to be the main emitters of GHGs and who therefore ought to bare greater responsibility. This necessitated a legal translation on the implementing rules of the Kyoto Protocol with the final accord proposing greenhouse gas emissions quotas with an option to resale of emission allowances between the developed countries and the G77. However, this is only one of many more controversies to dog climate negotiations.

A reading of the history of the Conference of the Parties dating back to the first United Nation Framework Convention on Climate Change (UNFCCC) COP 1 that took place in 1995 in Berlin, Germany, clearly reveals the main controversies that over the years took much of the oxygen out of international climate negotiations. Without going into details of each and every issue[5], there were signs as early as the first COP that the climate negotiations were going to be anything but agreeable. One of the contentious issues to emerge early on in the negotiations was joint implementation (JI) – a mechanism under Article 6 of the Kyoto Protocol allowing a country with an emission reduction or limitation commitment under the Kyoto Protocol (Annex B Party) to earn emission reduction units from an emission-reduction or emission removal project in another Annex B Party, each equivalent to one tonne of carbon dioxide (CO_2)[6].

Table 5.1: Conference of the Parties of the UNFCCC, when and where held (1995-2015) (Author's own compilation)

Conference of the Parties (COP)	Dates	Venue
1	March 28 - April 7, 1995	Berlin, Germany
2	July 8 - 19, 1996	Geneva, Switzerland
3	December 1 - 10, 1997	Kyoto, Japan
4	November 2 - 13, 1998	Buenos Aires, Argentina
5	October 25 - November 5, 1999	Bonn, Germany
6	November 13 - 24, 2000	The Hague, The Netherlands
7	October 29 - November 9, 2001	Marrakech, Morocco
8	October 23 - November 9, 2002	New Delhi, India
9	December 1 - 12, 2003	Milan, Italy
10	December 6 - 17, 2004	Buenos Aires,

---

[5] For a more comprehensive review of issues and outcome of every single COP held to date, visit *Climate Policy Observer: Monitoring Climate Policies,* hosted by the International Centre for Climate Governance. Available online: http://climateobserver.org/open-and-shut/carbon-emissions/

[6] UNFCCC, "Joint Implementation". http://unfccc.int/kyoto_protocol/mechanisms/joint_implementation/items/1674.php

		Argentina
11	November 28 - December 10, 2005	Montreal, Canada
12	November 6 - 17, 2006	Nairobi, Kenya
13	December 3 - 15, 2007	Bali, Indonesia
14	December 1 - 12, 2008	Poznan, Poland
15	December 7 18, 2009	Copenhagen, Denmark
16	November 29 - December 10, 2010	Cancún, Mexico
17	November 28 - December 9; 2011	Durban, South Africa
18	November 26 - December 7, 2012	Doha, Qatar
19	November 11 - 23, 2013	Warsaw, Poland
20	December 1 - 12, 2014	Lima, Peru
21	November 30 - December 12, 2015	Paris, France

'Trees for smoke'

In the build up to COP 1, majority of developing countries viewed JI as a means for Annex I Parties to avoid domestic action to meet their commitments under the Convention. Initial opposition to JI by developing countries was canvassed under the banner 'tree for smoke', which was informed by a variation in the position that JI include developing countries other than exclusively between Annex I Parties. Subsequent negotiations resulted in a shift and acceptance of participation of non-Annex I countries in JI on a pilot basis conditional to no credits accruing to any Party during the pilot phase. Annex I Parties grappled with the credits issue, however. Some countries, particularly the US, continued to insist on emissions credits during the pilot phase. Developing countries were also concerned that JI be supplemental and not substitute for funding and the financial mechanism established under the Convention.

The Ministerial Declaration to emerge out of COP 1 – the *Berlin Mandate* – was meant to operationalise the principle of "common but differentiated responsibilities" established in the UNFCCC. However, the agenda soon ran into headwinds at the COP 2 held in July 1996 in Geneva, Switzerland. Many disagreements on fundamental issues, such as adequacy of commitments intensified.

For example, delegates disagreed on references to the IPCC Second Assessment Report (SAR), especially the call for legally binding commitments. At the same time, the United States' sudden change of position to support a legally binding protocol, but one linked to a tradeable permit system, served to stalk, rather than calm, further controversy. There were concerns, for example, from developing countries concerning heavy reliance on market-based schemes, arguing that markets favour the wealthy and often solidify, rather than resolve, inequities.

The financialisation and commodification of carbon is a much rigorously debated issue.[7] The liquid global carbon market now stands at hundreds of billions of dollars. In 2015, six multilateral development banks jointly committed more than US$25 billion in climate finance. According to the 2015 Joint Report on Multilateral Development Banks' *Climate Finance*,[8] "in addition, a total of US$55.7 billion were committed in co-finance, raising the total to US$80.8 billion". Anne Petermann (2015) makes a similar point arguing that Reducing Emissions from Deforestation and Forest Degradation" initiative, or REDD+ is a scheme to "enable business-as-usual (BAU) under a green veneer".[9] By allowing polluters to buy forests rather than cut their emissions at source provides a convenient pathway for BAU disguised as real solution to the climate crisis driven by genuine desire to reduce emissions.

Pope Francis, in *his Encyclical Letter Laudato Si' on the Care for Our Common Home* given in Rome at Saint Peter's on 24 May, 2015 characterized market-centric strategies such as emissions trading schemes and offsets as "simply a ploy which permits maintaining

---

[7] Larry Lohmann, for example, argues that like financial derivatives, "carbon commodities work through a process of radical disembedding the climate issue from the historical question of how to organise for structural, long-term change capable of keeping remaining fossil fuels in the ground…Carbon markets aim at securing those background conditions for accumulation that are most dependent on fossil fuels and most threatened by calls for emission cuts" (Lohmann, 2012).

[8] http://www.adb.org/documents/joint-report-mdbs-climate-finance-2015. Accessed 13th August 2016

[9] Anne Petermann, 2015. "Confronting climate catastrophe: Direct action is the antidote for despair", *Pambazuka News* 753. Available online: http://www.pambazuka.org/land-environment/confronting-climate-catastrophe-direct-action-antidote-despair. Accessed 13th August 2016

the excessive consumption of some countries and sectors".[10] According to Patrick Bond (2015), the commodification of carbon sinks, especially forests but also agricultural land and oceans is route with contradictions.[11]

Despite the controversy surrounding its proposal and vehement objection from developing countries, the Ministerial Declaration adopted in the second COP reflected the United States position which rejected the scientific findings of climate change proffered by IPCC in the SAR. COP 2 also rejected uniform harmonized policies in favour of 'flexibility'. COP 3 was no less dramatic: Though the Kyoto Protocol (named after the Japanese City of Kyoto where the third Conference of Parties was held in December 1997) was eventually adopted after intense negotiations and whereas most industrialised national (Annex B countries) agreed to legally binding reductions of GHGs emissions, the US – the then leading polluting nation on earth – flatly refused to ratify the Protocol.

There was expectation that issues that were still outstanding from Kyoto would finally be resolved in Buenos Aires, Argentina, during COP 4. Such hope was however dashed in the face of persistent controversies and disagreements. Agreeing to disagree, Parties instead adopted the Buenos Aires Plan of Action under which they declared their determination to strengthen the implementation of the UNFCCC and prepare for the future entry into force of the Kyoto Protocol. The Fifth Conference of Parties was primarily a technical meeting and did not reach major conclusions. Summing the status during COP 5 held in Bonn, the German Chancellor Gerhard Schröder noted that, despite the establishment of the UNFCCC, there had been setbacks in the climate process, including the inability of most industrialised countries to reduce their CO_2 emissions to 1990 levels by the year 2000.

Further disagreement continued to mar negotiations over controversial proposal to allow credit for carbon "sinks" in forests and agricultural lands. Contradictions surrounding carbon trading

---

[10] Pope Francis, 2015 *Laudato Si,* Available online: http://w2.vatican.va/content/francesco/en/encyclicals/documents/papa-francesco_20150524_enciclica-laudato-si.html. Accessed on 13 August 2016

[11] Patrick Bond, 2015. "Carbon trading reborn in new-generation mega-polluters", *Pambazuka News* 753. Available online: http://www.pambazuka.org/governance/carbon-trading-reborn-new-generation-mega-polluters. Accessed 13th August 2016

led to suspension of COP 6 in The Hague. This was without agreement only to resume months later on. Bonn, with little progress having been made to resolve initial impasse on carbon trading as well as other contentious issues, included disagreements over consequences for non-compliance by countries that did not meet their emission reduction targets. There were difficulties in resolving how developing countries could obtain financial assistance to deal with adverse effects of climate change and meet their obligations to plan for measuring and possibly reducing greenhouse gas emissions. The follow up negotiations in Bonn, however, managed to reach agreement (the so-called 'Bonn Agreement') on flexible mechanisms, carbon sinks, compliance and financing setting the stage for COP 7, which was held in the Moroccan coastal city of Marrakesh from 29 October to 10 November 2001.

The outcome of COP 7 – the Marrakech Accord – was a compromise document that offered a degree of optimism about a future deal. However, it fell short of resolving outstanding technical issues relating to the structure of the Kyoto Protocol. This would have cleared the way for its entry into force. But, what does all this have to do with climate change policy-making in Zimbabwe? Well, to begin with the Kyoto Protocol was supposed to be the overarching framework and reference point for joint action by the Parties in addressing the climate crisis, all its limitations/weaknesses notwithstanding. Therefore, delay in reaching a deal on its enforcement weakened the legitimacy of the UNFCCC process. This in my view had two important implications for climate policy-making in Zimbabwe.

One, the slow progress in reaching a climate deal provided little incentives for a small nation like Zimbabwe with insignificant historical responsibility for anthropogenic climate change to deploy its scarce resources to developing a climate response framework and strategy. Two and partly related to the first point is the fact that by the time COP 7 was taking place, the 'Zimbabwean crisis'[12] was

---

[12] The 'Zimbabwean Crisis' here refers to the decade of economic turmoil which starting in November 1997 set of by the massive depreciation of the Zimbabwean dollar (>30% in a single fall) after an unbudgeted gratuity and pension pay out to the country's War Veterans was announced. A year before the onset of the crisis (1996) Zimbabwe's GDP of US$8.6 billion was the second largest of the 15-country Southern Africa Development Community (SADC), behind that of South Africa at US$143.7 billion. At the height of the economic

93

already underway meaning that government's hands were already full managing the crisis. It was therefore not to its interest to engage in additional policy arenas, including domesticating global governance framework such as the Kyoto Protocol that had proven to be mired in one controversy after another. I have referred to this phenomenon as 'a constricted policy space'. Under such a scenario, it is not surprising that the government of Zimbabwe made a strategic decision to engage in a selected range of policy issues it considered critical to its interest. It is probable that such a choice has consequences. If any, they are beyond the scope of this chapter.

Conclusion

In this essay I have argued that Zimbabwe's relatively slow pace in formulating its national climate change policy and response strategy is a reflection of the constricted policy space the country has had to operate in during the decade of 2000s till now. This forced the government of President Robert Mugabe to make choices between competing policy priorities. In this scheme of things, a national climate change was not among the government's topmost priorities. In addition, persistent controversies in the international climate change policy arena, conflagrated with perceptions of a 'regime change' agenda, sent the wrong signals resulting in less, not more, appetite for domestication of an emergent and problematic global climate policy architecture. These findings suggest that more rigorous analysis of the political economy of climate change policy-making is necessary to better explain why some countries tend to domesticate global policy frameworks faster, or better, than others.

References

AMCEN. (2011). *Addressing Climate Change Challenges in Africa: A Practical Guide towards Sustainable Development*. (African Union Commission and African Ministerial Conference on Environment, AMCEN).

---

paralysis in 2008, Zimbabwe's GDP reached only US$4.8 billion, falling to the rank of eleventh in SADC (Deve, 2012)

Bond, P. (2015). "Carbon trading reborn in new-generation mega-polluters", *Pambazuka News* 753. Available online: http://www.pambazuka.org/governance/carbon-trading-reborn-new-generation-mega-polluters. [Accessed 13th August 2016].

Carbon Trade Watch. (2015). "Introduction Special Issues". Available online: http://www.pambazuka.org/governance/introduction-special-issue. [Accessed on 13th August 2016]

CISDL. (2012). "The Principle of Common but Differentiated Responsibilities: Origin and Scope", *CISDL Legal Brief* (Montreal: Centre for International Sustainable Development Law).

Deve, T. (2012). "The underlying cause of Zimbabwean crisis", *Pambazuka News*. Available online: http://www.pambazuka.org/governance/underlying-cause-zimbabwean-crisis. [Accessed on 21 August 2016].

Lohmann, L. (2012). Financialization, commodification and carbon: the contradictions of neoliberal climate policy. *Socialist register*, 48(85), 90-107.

Marimira, C., and Odero, K. (2003). An analysis of institutional and organizational issues on Fast Track Resettlement: The case of Goromonzi. *Delivering Land and Securing Rural Livelihoods: Post Independence Land Reform and Resettlement in Zimbabwe. Harare and Madison: Centre for Applied Social Sciences and Land Tenure Center*, 259-68.

Meadows, D. H., Meadows, D. L., Randers, J., Behrens III, W. W. (1972). *The Limits to Growth: a report for the Club of Rome's project on the predicament of mankind.* New York: Universe Books.

Odero, K. (1999). "Local Authorities' Response to Restructuring". Department of Rural and Urban Planning Working Paper, Harare: University of Zimbabwe.

Odero, K. (2002). "What is the relevance of Regional Planning to Land Resettlement in Zimbabwe?" Paper presented at the Planning Africa 2002: Regenerating Africa through Planning Conference, Durban.

Odero, K. (2013). "The Impacts of Climate Change and Variability in Urban Settings", paper presented at the First Climate Science Symposium of Zimbabwe, 19-21 June 2013, Harare. Cresta Lodge.

Republic of Zimbabwe. (2013). Zimbabwe Second National
Communication to the United Nations Framework Convention
to Climate Change (Harare: Ministry of Environment and
Natural Resources Management).

Umemiya, C., Goodwin, J., and Gillenwater, M. (2016). "How did
intended nationally determined contributions (INDCs) address
capacity building? Implications for future communication by
Parties on capacity building under the Paris Agreement". *IGES
Working Paper.* (Institute for Global Environmental Strategies:
Kanagawa (July, 2016). Available online:
http://pub.iges.or.jp/modules/envirolib/view.php?docid=6691
. [Accessed on 08 June 2017]

Chapter 6

Urban Climate Resilience in Zimbabwe: A Review

Chipo Mutonhodza, Patience Mazanhi & Aurthur Chivambe

Introduction

One of the most perpetuating problematic issues threatening the world at large is climate change. This has however proved to be much fatal in the southern part of Africa particularly Zimbabwe despite all the efforts being done to curb the effects of climate change. Climate change has raised huge dilemmas for developing states such as Zimbabwe, as it is still a struggle to implement short term development goals and the mitigation of existing social, economic and environmental problems which may have to be balanced against the need for long term environmental and prospect of chronic climate change (Unganai, 1996). Global climate change has been associated with the greenhouse effect, ozone depletion and land surface changes which make it crucial for the country to anticipate and plan for increased future climate impacts (Unganai, 1996). In many of Zimbabwe's urban areas such Harare, climate changes have been triggered by the ever-increasing population growth rates and this has seen the deterioration of the natural environment as people compete for places to reside. As a result, anthropogenic heat releases during the night, a lot of vegetal cover which is being encroached and permanent hard surfaces have exacerbated temperature levels. Zimbabwe, like the rest of Africa, is constrained by its inability to put appropriate measures in place in order to respond to climate change requirements because of the lack of human, institutional and financial resources.

Urban heat islands have also increased, and this has been due to high incident short wave radiation at the surface, anthropogenic heat releases during the night and low soil moisture leading to high energy gain during the day (McCarthy 2010). When all these factors continue to take place, climate change effects will still continue to be detrimental. Although, the country might not want the negative effects of climate change, they still continue to affect the people due to the high levels of poverty and existence of already highly degraded environment and lack of background information on the direction and magnitude of climate change to anticipate. This, from statics as highlighted by (Unganai, 1996), have increased daytime temperatures over Zimbabwe by 0.8°C from 1933 to 1993 and precipitation declined by up to 10% on average over the period of 1900 to 1993.Therefore, it has had the highest effect over urban settlements. Desk research has been done through compilation of literature and review from Asia and Pacific whose experiences are rich in providing case studies and policy options that Zimbabwe can learn from. The chapter has been organised beginning with the framework of the debate and in this, the urban climate change effects are discussed noting the La Nina and El Nino phenomena. This is followed by the discussion on the urban climate resilience and as such, case studies from the Asia and Pacific have been used to get lessons. The settlement hierarchy of the country is discussed, the legislative, institutional and policy frameworks for climate resilience, the methods and techniques in enhancing urban climate resilience have also been explained as well as the case studies of the various cities and towns of the country have been used to strengthen the clarity of the discussion. The last passages of the chapter purport on the lessons learnt from the cases as well as the conclusion, policy options and the recommendations towards urban climate resilience.

Framing the Debate

The expansion of urban areas in Africa has been increasing recently due to a variety of factors. Urbanisation has enormously expanded leading to high population concentration in a few large cities especially capital cities. Nsiah-Gyabaah (2003) has defined urbanisation as the radical shift from a rural to an urban society and is an essential corollary of industrialization that goes hand in hand with the role of human scientific, sociocultural and technological development. This has been fuelled by the ever increasing rural to urban migration and high birth rates in the bigger cities. Africa's urban areas have expanded from 14.5% in 1950 to 34.5% and have also been projected to increase in 2025 by 4 times the 1990 rate (Nsiah-Gyabaah 2003). This is because of some beliefs that urban areas have much better employment opportunities, wages and social services hence High rural to urban migration. Due to the improvement of technology and services in many of Africa's cities key services such as health facilities are leading to low mortality rates and high birth rate. Much of development strategies are also biased towards urban areas and at the same time neglecting the rural side. However, even though urbanisation has many benefits to offer to the societies of Africa, it has led to the deterioration of the environment and has contributed to the climate change as there is increased competition for land and water (Nsiah-Gyabaah 2003). With the overpopulation resulting from this urbanisation, social services have proved to be insufficient. With the high-level demand for timber and fuel wood leading to inland degradation, loss of biodiversity, soil fertility decline and leaching of soil nutrients leading to the pollution of rivers. As a result of the above, climate change has resulted (Nsiah-Gyabaah 2003).

Climate change effects: *The El-Nino and La-Nina phenomena*

Climate change has many serious consequences towards the inhabitants of the earth. In many areas it has led to flooding sometimes due to the La-Nina effects and it has also caused droughts in many African areas. The issue of droughts has been common in the southern part of Africa and had been curtailed also by the encroachment of desert features in the areas. Shortage of water for both domestic and commercial uses has worsened leading to poor sanitation, poor service delivery and spread of many diseases such as cholera and typhoid. As defined by Unganai (1996),

Climate change refers to the shift of climatic conditions in a directional incremental mode, with values of climatic elements changing significantly and the evidences of it can be detected over several decades. Climate change, as highlighted by Taylor (2014), has led to gradual shifts in temperature, intense rainfall, rising sea levels, coastal erosion, ground water salinity and extreme events such as fires, floods, heat waves and storm surges. Due to the increased frequency and severity of heat due to climate change, a lot of diseases have been brought forth and this has affected the economies of many developing countries as additional costs of climate control over the environment may have to be met and the costs of curing diarrheal diseases and other infections (Satterthwaite, 2008). Climate change has been also involved with the La Nina and El Nino phenomenon. The WMO (2014) has defined El Nino, as the naturally occurring phenomenon involving

99

fluctuating ocean temperatures in the central and eastern equatorial pacific, coupled with changes in the atmosphere. La Nina has the cooling effect as it leads to the outpouring of heavy rains and floods. Therefore, with these deadly results of climate change, it becomes empirical for authorities to find better climate adaptation and resilience techniques in order to minimise negative impacts and harness chances and opportunities from the changing climatic conditions.

Urban Climate Resilience

Climate resilience has been seen as the only possible way of withstanding the negatives of climate change. Climate resilience is the ability of a social or ecological system to absorb disturbances while retaining the same basic structure and ways of functioning the capacity of self-organisation and the capacity to adapt to stress and change (Tyler and Moench, 2012). This calls for cooperative efforts to come up and implement better resilient strategies to minimise the evils of climatic changes especially in the urban premises. Urban climate resilience therefore is, as defined by (ACCCRN 2013 and Taylor and Peter 2014), is the capacity of cities to survive, adapt and think in the face of stress ad shocks and even transform when conditions require it by pro-actively adjust to reduce negative impacts and harness any new opportunities from changing climatic conditions, learning to navigate complexity and uncertainty across multiple scales. Urban climate resilience has proved to be the only way to go in adapting to the evils of climate change. Considering the negative effects of climate changes, urban climate resilience makes it possible to reach sustainability of the remaining environmental resources. Through urban climate resilience, communities become enlightened on how to properly manage and fend for themselves through adapting to climatic changes. Therefore, these enable them to not only withstand and resist climate circumstances but also to prevent further damage to the environment (Brown *et al.* 2012).

Many techniques can be used to achieve this urban climate resilience and the commonest of all is the involvement of the public in finding ways to resist the effects of climate change. A case study is that of Gorakhpur city in India which had been facing severe climatic change effects such as flooding, water logging, temperature extremes, power shortage, poor quality of water and increased incidence of water and vector borne diseases. In order to minimise the dangers, the public was made aware and they managed to improve drainage, housing, health and communication systems. Climate proofing was also done to the infrastructure of the city. All Therefore, was a success due to the advocacy and capacity building activities which were done to raise public awareness (Sharma *et al.* 2013). With stakeholder participation in climate resilience, the process of implementation is quickened, and it even motivates the public on the importance of managing the surroundings to protect themselves from much further evils which may result due to ignorance.

Case Studies of Asia and Pacific

Countries in the Asian and Pacific regions have managed to overcome a lot of climatic change challenges in many ways. This is effectively enhanced by the fact that many of such countries are now well economically developed and can easily fund for their proposed projects as compared to some developing countries such as Zimbabwe. One of the countries which have managed to counter some of the climatic change effects includes Brazil. In managing water stress as a result of the climatic changes, the state of the government has built huge environmental infrastructure thought public and private support hence creating opportunities for companies with water technology solutions in drought-stricken areas. Brazil's one of the most populated city, Sao Paulo, had been facing massive water challenges due to previous droughts, therefore this became an urgent matter. The state had to source from the international community to get new technologies in order to deal with the scenario, some of them include better irrigation methods and treating and recycling of industrial water for industrial and other uses. In order to achieve this, the government of Brazil made use of public-private partnerships. This allowed the inclusion of all stakeholders in curbing this problem. And as a result of such partnerships, large water utilities in Brazil such as SABESP and Odebrecht Ambiental formed an entity to provide drinking water and even recycled water for petrochemical companies and improved water technologies have been sorted to make the water problem minimised.

Some countries such as Ghana, Italy, Morocco and Rwanda have also made some adoptions in trying to curb the climate change

101

issues. As highlighted by (World Water 2015), the UK Company, Bowater managed to complete two major water projects whose sphere of influence is quite large. Italy also expanded its water portfolio in the city De Nora, in 2015; Morocco also completed an antipollution system hence preventing direct wastewater discharges into the sea. Lastly, in Rwanda, around six hundred thousand people had been facing water problems and as a solution they were given personal water backpacks that could make it easier for women and children to carry water from the source to home. Therefore, in analysing the success of implementations by other countries, it is thereby a call to many developing countries to copy and improve on some of the ways in arresting the evils of climate change. Of the above case studies, the main arm of success by such countries has been due to collaboration and involvement of many if not all stakeholders in the minimisation of climate change effects.

This has become imperative for Zimbabwe as a developing nation to copy and improve and even find better ways of urban climate resilience. Zimbabwe is a landlocked country situated in the southern part of Africa. The state comprises the savannah climate and on top of that desertification has been encroaching many of the parts of the country. Another contributing factor is the existence of the Kalahari Desert in Botswana which may somehow lead to desert symptoms in Zimbabwe. The country is in a semi-arid region as had been mentioned earlier, with limited and unreliable rainfall patterns and also temperature variations. The existence of more arid environments for agricultural production has also led to the shifting of Zimbabwe's five main agro-ecological zones or natural regions as rainfall patterns tend to decrease from region one to five. The country is bordered by Zambia to the North West, Mozambique to the east and Botswana to the South West and South Africa to the south. With the experiences of severe droughts in the country, there has been recently low agricultural output and also strained surface water (Brown et. am 2012).

Settlement Hierarchy

The settlement hierarchy in many developing countries of Africa particularly Zimbabwe, have been highly influenced by the colonisation of the country in past decades. Before the colonisation, the country constituted of scattered settlements without any planning form before. During the colonisation, urban units developed from military fortresses. After the country gained its independence from the whites, a one city concept was adopted. Urban settlements began to emerge. Urban settlements in Zimbabwe's context are a settlement comprising of about 2500 people or more and the majority of who do not partake in primary production activities and these may include agriculture and mining (Munzwa 2010). The morphology that is, the spatial form and structure of Zimbabwe's settlements may be ranked from the bottom to the highest settlement in terms of activities which take place in each settlement (Munzwa 2010, see Table 6.1)

Table 6.1: Settlement Hierarchy of Zimbabwe (Fieldwork, 2016)

Type of Settlement	Description
Business Centres	First and Lowest level of the hierarchy; very few shops privately owned
Rural Service Centres	Comes second on the hierarchy. These are headquarters for the village services through the WADCOS. There are similar activities to those of the business centres
District Service Centres	Similar and higher-level services to those mentioned above. There are coordinated and integrated services. Police stations also found in this stage
Growth Points	Follow the district centres and now provide tertiary services such as banking facilities
Towns	Include the above as well as the state and other private sector interventions
Cities	Comprise of all the above as well; these are the highest service deliverers and many decision-making comes from these

Business centres are at the lowest level of the settlement hierarchy of Zimbabwe's urban settlements. It constitutes of very few shops which may be five on average and these will be privately owned. As such, they offer daily services to their sphere of influence. Rural Service centre are the second from the business centres mentioned earlier. The activities do not highly differ from those of the above settlement. However, the rural service centres integrate functions of the lower order settlements such as villages and the main purpose of each service centre is to support a maximum of about 10000 people. These rural service centres operate as the headquarters for certain entities like WADCOs. They are also the providers of social services such as schools and clinics to their residents. Lastly, they are also responsible for the infrastructural developments such as roads and water supply as well as communication. From these, we move on to the third level of the settlement hierarchy which is the district service centre. This centre has similar and even higher-level services which comprise of the rural service centres and the district administration capitals. They coordinate and integrate services available at the other lower order entities in the district. Some additional services which are in accordance with the district requirements such infrastructure developments of hospitals, administration offices, police stations and even information centres are provided.

Growth points follow after the district service centres and have also similar activities. However, these now provide services of the tertiary sector such banking. These growth points are also large even in terms of area of coverage.

Towns as centres comprise all of the above and also include the government and the private enterprises. The last and one on the apex of the hierarchy is the city. The morphology of cities has evolved from concurring with the pattern of primacy. Primacy is the fact of being the most important person or thing (Oxford Dictionary). This means that cities have grown to become the central harbour of many activities and has also been considered as the most important of the hierarchy. However, concerning the issue of climate change, cities have become the quandary of environmental challenges which include pollution on the land, air and water. Also, due to the degradation of resources as ruralisation has created into the city. Since they are the centrifugal of all activities, some problems such as space contestation, siltation of water bodies like dams and this has greatly disturbed the ecosphere (Munzwa 2010).

Legislative, institutional and policy frameworks for climate resilience

Zimbabwe has made so much progress in finding ways of adapting to climate changes, however the implementation may have been slowed down due the declining economic conditions as funds to make the implementation necessary would not be readily available. In a bid to find ways of doctoring the climatic change effects, the country introduced environmental management boards which enable the protection of the environment. This has seen the formation of the environmental management agency board in 2002 as according to the Environmental Management Act no13 which concerns itself with the management of natural resources and the protection of the environment (Chagutah, 2010).

Most climate change adaptation responsibilities are implied in many incongruent sectoral responsibilities and the opportunity for adaptation planning is given in sections 87and 97 of the Environmental Management Act which require the heads and local authorities to come up with a national environmental action plan and local environmental action points. Some of the legislative tools introduced include the national water act no31 of 1998 and the Zimbabwe water authority act no11 of 1998 also which were promulgated after the 1991 to 1992 drought. This was in a bid find ways to minimise the harsh effects of drought as water shortage had become rampant. The other climate change management policy act adopted is the disaster management policy from the civil act chapter 10.06 and it focuses mainly disaster response than comprehensive disaster management (Chagutah, 2010). However, from the current outlook of the current situation in the country, there is no a comprehensive, blunt national policy and legislative framework for climate change and its adaptation. This is because, the legislative and programmatic adaptation responses are found in a plethora of development strategies and action plans of the various government sectors such as the environment and natural resources management, agriculture and disaster management sectors and it is through these broad range ministries that the climate change adaptation are found (Chagutah 2010). Therefore, it makes it difficult to fully implement some of the strategies brought forward towards the ratification of climatic changes since there is no clear-cut rule of thumb or national strategy to implement the provisions. Of late, the irrigation farmers, as highlighted by Chagutah (2010), are at a greater risk from the increase in the frequency of droughts since many of such farmers are less diversified. It therefore calls for better methods by the state of Zimbabwe to upgrade the climate change resilience strategies.

Methods and Techniques in Enhancing Urban Climate Resilience

There are many methods and techniques which have been put forward to curb the dangers and of climatic changes and to strengthen the resilience strategies to this climate change situation.

In Zimbabwe, much of has been done by the environmental management agency board in coping with drought and climate change in the country. The University of Zimbabwe and the soil fertility consortium for southern Africa have tried to explore the measures that can enhance the capacity of local communities to the pressures of climate change. The midlands state university also carried a study concerning the building of capacity to adapt to climate change both in Zimbabwe and one of its neighbouring countries, Zambia. However, despite the willingness to implement such projects, only two were manageably completed (Chagutah 2010). In making these techniques a success, there is a strong need to advance the development of the adaptation strategies in the biodiversity and natural resources management sectors (Perez et. al). Some of the challenges and constraints emanating in trying to create resilient strategies that include the issues of uncertainty in the way climate change may directly and indirectly impact agricultural and food systems.

Case Studies

Harare

Harare is the capital city of Zimbabwe falling in Region 2 and many of the residents of the city obtain their water local boreholes yet much of the water obtained is highly contaminated with faecal matter (Musemwa, 2010 and Nhapi *et al.* 2006). Also, it is even difficult to obtain water from these boreholes since they seem not to be proportional to the number of households depending on them, hence, leading to very long queues and spending of the whole night in a queue waiting for one's turn to get the water. Due to this, many are forced to buy from water sellers as it will be best alternative as even many of the boreholes are not functioning very well due to overuse and high congestion on them (Manjengwa *et al.* 2014).

The city has been facing many environmental challenges and this has affected even the agricultural base of the city. The recent extreme climatic conditions of droughts and cyclones with the much exacerbating population growth rates has led to the channelling of the already few resources of the country towards the provision of better and cleaner water resources and sanitation services (Nhapi, Siebel and Gijzen, 2006). The city has been facing very serious water management problems as it drains from Lake

106

Chivero and also depends upon it for domestic and industrial water needs even though the Lake is highly polluted, and the pollutants not effectively removed through wash water. The lake has been reportedly said that its water table has fallen giving rise to the shortage of water in the city (Nhapi, Siebel and Gijzen 2006). The other problem in the city has been that of food shortages which has seen many children getting married earlier than normal. This has been in order to sustain themselves from the hunger especially in the areas such as Mabvuku, Tafara, Glenview and Mufakose. These areas are said to be the districts with the largest proportion of people who have witnessed the floods and high rainfall conditions hence poor agricultural results from their urban farms (Manjengwa *et al.* 2014). However, not much progress has been put in place to produce the national climate change strategies (Shumba and Carlson, 2011).

When it comes to climate change, Zimbabwe signed and ratified the UNFCCC in 1992 (Roberts, 2008). It has since submitted two national communications, in 1998 and 2013. The government has renamed the previous Ministry of Environment and Natural Resource Management to the Ministry of Environment, Water and Climate and is developing a National Climate Change Response Strategy. The draft strategy of 2013 includes a national action plan for adaptation and mitigation, analysis of strategy enablers, climate change governance and implementation framework. Climate change considerations have also been incorporated into some aspects of development and growth planning, like the Medium-Term Plan (MTP) of 2011 -2015.

Mutare

Mutare is the fourth largest city in the country situated in the Manicaland Province in Eastern part of the country. It is also in Region 1 of the country and has a temperate climate with an average of 19C temperature and 818mm annual average rainfall. The city experienced flash floods which affected some residential areas such as the Zimta Park and the Dream house because of the heavy downpours in 2014 which led to the destruction of a lot of property, houses and other valuable assets. The areas have poor drainage which worsened the situation. However, the Zimbabwe Red Cross Society came to the rescue of the flood victims (ZRCS, 2013). Mutare has also been facing drought related problems which saw a lot of people migrating from the drought prone areas to seek

107

greener pastures in the less affected areas. This has been highly prevalent in the eastern highlands where illegal homes built of wood, mud and thatch have been increasing as illegal settlements. The drought has been said to be related to the climatic changes occurring in the whole world at large. This has also exacerbated the poverty levels of many people as they even fail to grow enough of the staple food, maize. In sight of that, about $4 million was availed towards the reclamation and rehabilitation of dams such as the Osborne Dam, irrigation schemes, rainwater harvesting and borehole and well drilling (*NewsDay*, The, 2015). With the ongoing climate change, soon the areas in the region 1 of the country will suffer the effects of such climatic changes in the future. Therefore, it is of importance that the planning authorities of the city take immediate action in finding better and more sustainable ways of facing the detrimental effects of climate change.

Masvingo

Masvingo is one of the largest provinces in Zimbabwe and the one most experiencing fluctuating rainfall patterns due to the climatic change effects and this has become a huge problem to the province as a whole. According to Mawawa (2016), Masvingo is a perennial drought zone area. Because of these occurrences, many of the people in Masvingo lost much of their valuable assets and wealth such as cattle and donkeys as many of these animals died because of the shortage of water and enough pastures for grazing. Many of the people were also affected due to poor harvests as the crops failed because of the lack of enough rainwater Masvingo experienced an unexpected flooding of its kind in the year 2014. The unexpected heavy rains led to the over flooding of the dam leading to the collapse of the Dam Wall. A lot of houses and property of poor and innocent families were swept away, to that, relief came as the state deployed about 20 vehicles to evacuate the victims immediately and the Red Cross Society also offered tents to their aid (Herald, The, 2014).

Victoria Falls

Much of the water in Victoria Falls is from the mighty Zambezi River in the North-Western part of Zimbabwe. The highest flow recorded was experienced in 1957. Latter it was due to the El Nino events (Tumbare, 2000). Sometimes the floods occur due to overflows. The area has also experienced heat waves in the past years. The effects of climate change on human health continue to be a matter of scientific debate. Some assessments suggest that climate change can lead to an expansion of the areas suitable for malaria transmission. In the case of Victoria Falls, malaria transmission is more likely to increase than decrease in the central plateau where the population is currently concentrated. However, it is noted that non-climatic factors will also influence the future geographic distribution of malaria, including parasite drug resistance, demographic change and changes in land-use patterns.

Climate change is affecting not only Victoria Falls in Zimbabwe but its neighbouring countries as well. In the framework of the Southern African Development Community (SADC) they make a concerted effort to find solutions. In addition, there are networks on the continent that exchange information about climate change. There is also regional cooperation in climatic and environmental research, for example on the areas along the border river Limpopo and Zambezi. Despite a law on environmental management having been passed in 2002, there is no comprehensive climate policy or national adaptation strategy in Zimbabwe. On a local level in Victoria Falls, people are developing their own forms of adaptation to the changes in climate. The agro-ecological knowledge, gathered mainly by women over several generations, has acquired a new level of significance.

Gwanda

Gwanda is a small town situated in the southern part of the country between Beitbridge and Bulawayo. The town experiences temperatures similar to that of Beitbridge and some problems of water shortage. The residents of Gwanda especially the villagers in the outskirts of the town have been seriously hit by hunger due to the unexpected drought occurrences in the area. However, some drought relief aid was given to the victims of the drought (Chiwanga, 2016). Flooding has been experienced in the area which led to the overflowing of the Mtshabezi River Bridge. This was severe to the extent of blocking even the old Gwanda to Bulawayo road and has also led to the spilling of the Mtshabezi Dam (Bulawayo 24 News, 2014).

Climate change, which is regarded as one of the biggest threats to the people's livelihood and development in Gwanda, is foreseen to have large impacts on livelihoods through impacts on agriculture and water resources. This include increasing occurrence of crop failures, pests, crop disease and the degradation of land and water resources (Khomo, 2006 and Chatiza *et al.* 2011). Gwanda poorer segments of society will be disproportionately affected, as mainly the people living in poverty and extreme poverty are communal farmers. Recent vulnerability assessments show that areas regarded 'excellent' for maize will decrease from the current 75% to 55% by 2080 under the worst-case scenario (Musarurwa, 2012). Over five million Zimbabweans (almost a third of the total population of 14 million) live in semi-arid zones and will suffer disproportionately from the emerging impacts of climate change and variability. These include natural disasters associated with extreme weather events such as droughts, periodic flooding, disease outbreaks for both human and livestock and loss of crop lands. Research shows that climate change has already caused a shift in Gwanda agro-ecological zones. Soil conditions and crop variety will be important aspects in minimising the losses.

Chiredzi

Chiredzi is an area located in the regions 4 and 5 and is one of the places in Zimbabwe which incur very high temperatures. The town of Chiredzi, due to this, has been frequently facing food shortages and water scarcity mainly during the drought seasons. The residents of the district also sometimes face urinary health problems due to that. Many of the household dwellings in the district are made up of pole and dagga which is not durable especially in the extremely bad weather events. Many of the people in the district face severe water challenges as they receive water from unprotected wells and boreholes. Due to the ongoing climatic changes, the rain seasons now experienced are short and cannot sustain the life and growth of many plants in the farms of the residents. As a result, the people have been facing food challenges leading to many girls getting married for a few bags of maize and other food substances (Manjengwa, 2014). Farmers in the region have also been facing decision problems as to how to allocate the limited resources and make the most out of the limited amounts of water as this has also some adverse effects on the welfare of the people (UNDP, 2012).

Community-based adaptation to climate change is one such approach and is increasingly widely adopted in Zimbabwe. It recognises that climate change impacts will fall hardest on those who are least able to cope, that responses will require local adaptation planning and a greater focus on building adaptive

110

capacity and that individuals and communities already have a strong reservoir of skills and knowledge that could increase their resilience (Sango, 2014). The vast majority of Zimbabwe's community-based adaptation projects target smallholder farmers in rural areas. A particularly good example, the Coping with Drought and Climate Change in Chiredzi District project, demonstrates how community based adaptation can empower local farmers, ensuring they actively participate in developing culturally sensitive and locally appropriate adaptation strategies for future climatic changes (Chatiza *et al.* 2011).

Box 6.1: Coping with drought and climate change in Chiredzi District
(Chatiza *et al.* 2011)

This five-year pilot project (2007–12), led by the Government of Zimbabwe, the United Nations Development Programme and the Global Environmental Facility used a community-based adaptation approach to assess vulnerability and to develop priority adaptation strategies for smallholder farmers and pastoralists in Chiredzi District. The focus was on food security and sustainable natural resources management. The project followed a fie step approach:

1) Assessing current and future climate risks and identifying those that the smallholder farmers considered most important.
2) Assessing the vulnerability of people's livelihood systems.
3) Identifying and assessing vulnerable communities (finding 'hotspots').
4) Discussing with the community to identify priority adaptation strategies.
5) Implementing pilot projects in 'hotspots'.

Importantly, this pilot project was a partnership between national government and civil society that aimed to up-scale local adaptation lessons towards national-level policy. This 'linked-up' policy approach is relatively new in Zimbabwe and could greatly help the country develop its dedicated national climate change framework.

Beitbridge

Beitbridge is a border town situated in the Southern part of Zimbabwe with the Limpopo River bordering the country with South Africa and with an area of coverage of 12697km2. According to the 2002 census report, the town had about 104000 people and the numbers have been increasing over the past years as many people tend to move towards South Africa. The town is one of the areas in the country which experience very high temperatures especially during the summer seasons and flooding and heavy storms in the rainy season even though infrequent rains are received

111

in the town. The Bureau (2016) highlights flash floods which almost swept nearly 780 individuals in March of the year. The vegetation is also very sparse since most of the characteristics defining the town are desert-like features. Also, The Town has been facing many environmental challenges on the Limpopo River and some other health problems in water resources management. In 2005, the municipalities of Beitbridge and Musina made an agreement to establish a partnership in the management of the environmental resources with committees sub-divided into focusing on tourism and conservation, disaster management, safety and security and lastly, environment and engineering. The main thrust of the partnership was to find the most profound ways of managing the Limpopo River as it is the source of water for both small towns. The river has been facing much pollution from the industries and residential areas as both the liquid and solid wastes are discharged. Also, a lot of water abstractions are done in the river as many people tend to fetch water for domestic and construction uses. However, huge efforts have been put in place to minimise water loss in the river and to also reduce pollution levels and to limit water demand while at the same time optimising the usage of water within the basin (UNEP, 2007). The climatic conditions of the town go to the extremes as the temperatures have made it difficult for the residents to even practice urban agriculture hence relying on the produce from Harare. Also, the water from the boreholes is salty and makes it very difficult to perform household chores which require the use of water mainly like cooking, washing and even drinking.

Table 6.2: Urban Centres and Climate Change

Urban Centre	Climate-Related Events	Level of Preparedness	Measures in Place	Remarks
Harare	-Flooding in the C.B.D during the rainy season -very high temperatures during the rainy season -lowering of water tables of the city's water sources e.g. lakes which supply the city with water -the urban heat island in the city's -very few vegetation in much of the city's areas	-very low -not many climate change strategies have been put in place (Shumba and Carlson, 2011). -reluctance on the preservation of wetlands	Boreholes drilled by NGOs during the cholera period of 2008 -EMA enforcement of laws governing the environment -introduction of blend fuels to minimise air pollution and the enforcing of the Polluter Pays Principle in order to reduce ozone layer depletion -tree planting	-there is much effort needed to revive the efforts of the city in meeting the challenges of climate change as well as intensively -Develop mechanisms for sharing information and collaborating with different actors, including civil society, the private sector and government at all levels.
Mutare	-flash floods -drought occurrences in the region		-Irrigation -afforestation -legislation -cloud seeding	- Climate change is also prompting a growing number of NGOs and research organisations, including UN agencies, to build strong adaptation components into new development projects, particularly in drought-prone rural areas.

113

Gwanda	-droughts -flooding of the Mtshabezi River	-very low	-Drought relief aid schemes -drainage and bridge renovations -cloud seeding	-Actively seek out and include marginalised groups in more participatory and inclusive decision-making processes and planning procedures to ensure that future climate policy responds to the vulnerabilities of both women and men living in climate sensitive areas of Gwanda.
Masvingo	-severe droughts (perennial) -flooding	-very low due to lack of funding		-Develop mechanisms for sharing information and collaborating with different actors, including civil society, the private sector and government at all levels. -Land-use planning tools are particularly important to reduce vulnerability to floods and other extreme-weather events.
Victoria Falls	-high rainfalls experienced also as a result of the waterfalls. -very high temperatures during the dry season	-sufficient since there is proper management of the Zambezi River's catchment area and capacity	-River catchment management	

Chiredzi	-increased drought incidences -reduced rainfall and shortening of rainy seasons -declining water bodies -flooding -annual temperatures projected to increase across the district	-Irrigation -afforestation - Community-based adaptation to climate change is one such approach and is increasingly widely adopted in Chiredzi especially by farmers.	-Improve how community-based adaptation and other development projects use information on climate trends and changes — this can help ensure that development benefits are not lost to climate change.
Bulawayo	-heavy torrential rains leading to floods -high temperatures and the encroachment of desert like features in some of the areas in the regions		- More attention should be paid to the multi-level nature of policy and programming

115

Lessons Learnt from the Cases

It is therefore imperative that many of the adoptive lessons from the cities be taken into account. Harare has been neglecting on the implementation of some of the environmental management procedures. The Issue of poor implementation has been reasoned with as being caused by lack of finance; however, this might not be the issue. Tree planting is also another lesson and it is being encouraged by the EMA and also done by the Nyaradzo Funeral Company. From the Victoria Falls management of the catchment area of the Zambezi River, it is very important that the planning authorities to adopt that in managing the other rivers in a bid to minimise the climatic change effects hence reducing or minimising the problem of water shortages. Also, the drilling on of boreholes is another lesson one can learn from these cases and it is also important that wells be made at household levels such that families and other individual homesteads be able to get enough water in the case of water shortages due to the normally prevailing drought conditions in the country. It is also important to involve the issue of community-based adaptation in order anticipate, cope with, respond to and recover from climatic variability and extremes (UNFCCC, 2007). Also, a growing number of the Non-Governmental Organisations and some other research institutes like UN Agencies are engaging themselves in the various societal developments and other projects (Brown *et al.* 2012). However, of all the lessons learnt from the above cases, there is need for community engagement since many of the responsive solutions to climate change effects lacked much consultation of the stakeholders save for Beitbridge as it has engaged in the management of the Limpopo River with the Musina Town Council. Climate change is a new subject in Zimbabwe. Therefore, developing a practical national response programme requires skills in many technical and social disciplines. Expertise in research related to climatic change is important. Prior to the eminence of climate change debate, Zimbabwe had developed research capacity mainly in agriculture, mining and various social disciplines but remains ill-equipped to conduct industrial research with assessment of technologies for industrial mitigation options.

Overall, due to lack of financial and human expertise, Zimbabwe has no basic climate change research programmes. However, research in climate change-related issues is carried out by

Government institutions in the areas of agriculture, water resources, energy and forestry, among others. The Southern Centre for Energy and Environment (SCEE), a non-governmental organisation in Zimbabwe, carries out climate change research for possible mitigation options in industry, particularly in energy efficiency.

Conclusion, Policy Options and Recommendations

Climate change has become the talk of almost all conversations as it has become the most prevalent detrimental factor in the environmental sphere. Therefore, it is very important that many legislative tools be put in place to both stop the change and to minimise the bad side of the effects of the climate change. As propounded by Stringer (2009), the UNFCCC expects that, of the 49 least developed countries of the world, of which 33 are in Africa, to partake in the National Adaptation Programs of Action (NAPAS). These are policy frameworks meant to identify and conscientise of the critical important adaptation procedures for further procrastination might increase the resistance and vulnerability or might lead to an increase in the costs in the long run (Stringer *et al.* 2009). In response to that legislative measure, as Zimbabwe also falls in the category of the least developed countries, after considering its lower gross domestic product and the higher levels of poverty in the country, it has taken also some routes to partake in that. The country has an environmental management board (EMA) and also the Zimbabwe Environmental Law Association (ZELA), which promotes the inclusion of the public and other stakeholders in the participation towards climate change governance (Brown *et al.* 2012). Therefore, it is important that in managing the environment and other climatic change effects, to involve all stakeholders in the planning and implementation.

Intensive and extensive educational campaigns must be carried out so that behaviour change might be instilled in the people. So many times, the majority always complains and blame the economy to the failure of the implementation of plans due to the current financial stability. However, to some extent it is because the people's minds are terrible and not the economy as such. Because of such vulnerability on the part of the people, even when educated on the importance of managing the environment in order to minimise the climate change problems, the people are still found cutting down trees and even practicing poor farming methods.

117

Also, there is need for gender balanced planning in the promotion of climate resilience. Women must be engaged in every aspect t as mostly they are the ones involved in the fetching of water and firewood hence are the ones mainly and directly impacted by these detrimental issues (UNFCCC 2007). It is now very important that the planning authorities reconsider and reconceptualise the issue of participation in all planning projects and that climate change information be harnessed into all possible projects. There is need to grab the opportunities of being a least developed country, Zimbabwe must take advantage and explore new adaptation funds such as the UNFCCC Adaptation Fund, hence making the problem of "there is no money" issue of a past history (Brown *et al.* 2012). According to the UNFCCC (2007), there is need for exploring new adaptation techniques in meeting the climate change effects. It further defines adaptation as a "process through which societies increase their ability to cope with the uncertain future, which involves taking appropriate action and making the adjustments and changes to reduce the negative impacts of climate change" (UNFCCC, 2007). Any future climate change policies in Zimbabwe will have to take into cognisance energy sources and their distribution as well as the differential socio-economic strata of the Zimbabwean population.

References

ACCCRN. (2013). Urban Climate Change Resilience in Action: Lessons for Projects in 10 ACCCRN Cities. Rockefeller Foundation,

Brown, D., Chanakira, R. R., Chatiza, K., Dhliwayo, M., Dodman, D., Masiiwa, M., and Zvigadza, S. (2012). *Climate change impacts, vulnerability and adaptation in Zimbabwe* (pp. 1-40). London, UK: International Institute for Environment and Development.

Chagutah, T. (2010). *Vulnerability and Adaptation Preparedness in Southern Africa*. Cape Town, Heinrich Boll Stiftung Publishing.

Chatiza, K., Brown, D., and Chanakira, R. R. (2011). *Climate change impacts, vulnerability and adaptation in Zimbabwe*. London, International Institute for Environment and Development.

Chiwanga, S., (19/03/2016). Drought Relief Maize Arrives in Gwanda *The Chronicle*.

Herald. 100 Families Evacuated from Flood Hit Suburbs in Bulawayo. S. Huru, Available online: https://www.herald.co.zw/100-families-evacuated-from-flood-hit-suburbs-in-bulawayo/[Accessed on 16 February 2018]

Herald. (5/2/14). Floods Wreak Havoc in Masvingo. Available online: https://www.herald.co.zw/floods-wreak-havoc-in-masvingo/ [Accessed on 16 February 2018]

Manjengwa, J. M., and Matema, C. (2014). *Children and climate change in Zimbabwe.* Harare, University of Zimbabwe.

McCarthy, M. P., Best, M. J., and Betts, R. A. (2010). Climate change in cities due to global warming and urban effects. *Geophysical Research Letters, 37*(9), 1-15.

Munzwa. K. (2010). Urban Development in Zimbabwe: A Human Settlement Perspective. Harare, University of Zimbabwe.

Musarurwa, C., and Lunga, W. (2012). Climate change mitigation and adaptation: Threats and challenges to livelihoods in Zimbabwe. *Asian Journal of Social Sciences and Humanities,* 1(2), 25-32.

Musemwa, M. (2010). From 'sunshine city' to a landscape of disaster: The politics of water, sanitation and disease in Harare, Zimbabwe, 1980–2009. *Journal of Developing Societies,* 26(2), 165-206.

Newsday. (1/09/2015). Fleeing Drought, Climate Migrants Press Zimbabwe's Fertile East.

Nhapi, I., Siebel, M. A., and Gijzen, H. J. (2006). A proposal for managing wastewater in Harare, Zimbabwe. *Water and Environment Journal,* 20(2), 101-108.

Nkomo, J. C., Nyong, A. O., and Kulindwa, K. (2006). The impacts of climate change in Africa. *Final draft paper submitted to The Stern Review on the Economics of Climate Change.*

Perez. A. A., Fernandez. B. H., Andgatti, R. C. (2010). Building Resilience to Climate Change, Ecosystem Based Adaptation and Lessons from The Field. IUCN, Gland.

Roberts, D. (2008). Thinking globally, acting locally institutionalizing climate change at the local government level in Durban, South Africa. *Environment and Urbanisation,* 20(2), 521-537.

Sango, I. (2014). *An investigation of communal farmer's livelihoods and climate change challenges and opportunities in Makonde rural district in Zimbabwe* (Doctoral dissertation).

Saroch, E., Mustafa, D., Ahmed, S., and Bell, H. (2011). Climate Change and Urbanisation: Building Resilience in the Urban Water Sector. A Case Study of Indore. ISET, Boulder Company and Pacific Institute, Oakland.

Satterthwaite, D. (2008, January). Climate change and urbanization: Effects and implications for urban governance. In *United Nations Expert Group meeting on population distribution, urbanization, internal migration and development* (pp. 21-23). New York, DESA.

Sharma, D., Singh, R., and Singh, R. (2013). Urban Climate Resilience: A Review of the Methodologies Adopted Under ACCCRN Initiative in Indian Cities. IIED Publishing.

Shumba, E., and Carlson, A. (2011). Status of and Response to Climate Change in Southern Africa: Case Studies in Malawi, Zambia and Zimbabwe. W.W.F, Harare.

Taylor, A., and Peter, C. (2014). Strengthening Climate Resilience in African Cities: A Framework for Working with Formality. African Centre for Cities, Cape Town.

The Bulawayo 24 News. (2014). 7 Feb, Mtshabezi River Maroons Gwanda Town Again.

Tumbare, M. T. (2000). Mitigating Floods in South Africa. Harare, Zambezi River Authority.

Tyler, S., and Moench, M. (2012). A framework for urban climate resilience. *Climate and development*, 4(4), 311-326.

.U.N.D.P. (2012). Coping with Drought and Climate Change in Zimbabwe. Available online: https://www.adaptation-undp.org/projects/sccf-cwdcc-zimbabwe. [Accessed on 17 June 2016]

UNEP. (2007). Tale of Two Cities: *Partnerships for Urban Sustainability*. UNEP and UN-HABITAT, Nairobi.

UNFCCC. (2007) Climate Change Impacts, Vulnerabilities and Adaptations in Developing Countries. UNFCCC.2009. Least Developed Countries under the Unfccc.

Unganai, L. S. (1996). Historic and future climatic change in Zimbabwe. *Climate research*, 6(2), 137-145.

World Water. (2015). Brazil: Opportunities and Challenges in Water Crisis. WEF Publishing, UK

ZRCS, (18/05/2013). ZRCS to Enhance Financial Management, Available online: http://projects.worldbank.org/P133424/public-financial-management improvement-consolidation-project?l ang=en. [Accessed on 17 June 2016]

Chapter 7

Climate Change Governance and Physical Planning in Zimbabwe

Liaison Mukarwi & Wendy Tsoriyo

Introduction and Background

Across the globe, climate change intensifies the hazards that affect the social systems and weakens resilience of states in tackling uncertainty and disasters (O'Brien *et al.* 2006). It also contributes to increased climate extremes (droughts, heat waves, cyclones etc.) and exacerbates adverse impacts (extreme hunger and poverty) (Birkmann and Mechler, 2015). Zimbabwe, like many countries, is experiencing drastic effects of climate change hence the need to prioritise the climate change reduction, mitigation and adaptation alike. The incoming of Sustainable Development Goals (SDGs) particularly, SDG Number 13 which speaks to the need to take urgent action to combat climate change and its impacts, among other global efforts also affirms the global world's commitment to combat and adapt to climate change. To this end, Zimbabwe signed the Paris Climate Change Agreement and pledged its commitment to mitigate and adapt the phenomenon in line with SDGs. Zimbabwe has also set a target of reducing greenhouse gas emissions by 33 percent as a way of operationalizing and walking the talk on climate change (Gukurume, 2013; Rankomise, 2015). Other efforts include the drafting of the National Climate Change Response Strategy (NCCRS) in 2013 which stated the national action plan for adaptation and mitigation, analysis of strategy enablers (e.g. capacity building, climate change education, communication and awareness), climate change governance and implementation framework (IEED, 2013). Also, the Environmental Management Act established environment related agencies and provisions aiming to provide for the sustainable management of natural resources and protection of the environment and the prevention of pollution and environmental degradation (Gukurume, 2013).

In this light, some may say Zimbabwe has done much in climate change mitigation and adaptation. However, it is sad to note that these efforts are not only compounded but also ineffective in the current governance structure (Gukurume, 2013; Mkandla, 2014). The governance framework has not been so integrated and holistic; many players and authorities are involved in the management of the environment stationed in different departments of the government and non-state actors (Mkandla, 2014). Of the players, physical planning has not been adequately empowered and equipped to influence the whole process. Physical planning is critical in translating ideas into reality through work programmes and action plans hence a very important component in climate change mitigation and adaptation processes (Economic Commission for Europe, 2008). Due to this obtaining, initiatives and innovations in climate change adaptation (which always have a physical planning essence) by various stakeholders seem to remain unimplemented (Mkandla, 2014; Chirisa *et al.* 2016). There is limited coordination by the Ministry of Environment, Water and Climate and the Department of Physical Planning and the Civil Protection Unit, all under the Ministry of Local Government, Public Works and National Housing. This represents a climate change governance problem which poses challenges in climate change adaptation in Zimbabwe and abroad (Chirisa *et al.* 2016). Therefore, climate change governance institutions must be reformed and adapted to handle emerging issues of climate change mitigation and adaptation (Gukurume, 2013; Rankomise, 2015). This therefore warrants this study as a tool to make various players can meaningful contribute to the debate and practice of climate change adaptation in the country. In this view, this chapter is structured to include the following subheadings Introduction and Background, theoretical framework, methodology, results and discussion, policy options and recommendations and conclusion.

Theoretical Perspectives

In a political ecology and environmental policy, climate change governance is the diplomacy, mechanisms and response measures aimed at steering social systems towards preventing, mitigating or dating to the risks posed by climate change (Koch *et al.* 2007; Forino *et al.* 2015). Regarding climate change governance, issues to be considered include, the state of adaptation preparedness;

institutional arrangements and capacities; the scale of funding required for adaptation; the best ways to administer development cooperation support; effective mechanisms for delivery; and mechanisms to ensure that adaptation efforts target and benefit the most vulnerable sectors of society (Birkmann and Mechler, 2015). This indicates that climate change mitigation and adaptation affect multiple sectors (multi-sectorial) and cannot be placed in one department or organ of the state calling for a multi-stakeholder approach (*ibid.*). These therefore mean that climate change governance requires a coordinated approach.

Notion of Complexity and the Climate Change Adaptation Debate
The complexity of the sustainability/climate change planning and decision-making processes are challenging policy-makers given that it has evolved to be a mainstream political issue (Giddens, 2008: 3). Politics involves making common decisions for a group of people; it is the activity by which differing interests within a given unit of rule are conciliated by giving them a share in power in proportion to their importance to the welfare and survival of the whole community. The "politics of climate change" is one of the key elements shaping current and future responses to this global challenge (*ibid.*). Politics of climate is usually presented as global politics (related to the contested negotiations of the United Nations Framework Convention on Climate Change (UNFCCC). as national politics (for example, the tension between the Obama White House and the US Congress over climate change legislation), or in the role of global corporate actors in the international politics and economy of climate change (Tanner and Allouche, 2011). Because there is no substantive framework for policy which offers coherence and consistency as to how national governments should cope with the long-term political challenges of climate change, this notion of politics has compounded coordinated efforts in fighting climate change and increased the complexity (Hall and Murphy, 2012a).

Conflicts of power and interest are inevitable in relation to climate. The conflicting views on the primacy of sustainability/climate change policy drivers can become polemic, delaying action on either front (Giddens, 2008; Sweeney *et al.* 2013). The influence of groups that fear adverse consequences of mitigation policies combined with scientific uncertainty, the complexity of reaching global agreements, the natural tendency for governments to delay action, has increased the complexity in climate change adaptation (Sweeney *et al.* 2013). Climate change mitigation or adaptation involves altering the way things are being done today especially in terms of production and consumption practices in key sectors such as energy, agriculture and transportation. The obtaining conflicting views on the change of lifestyles and livelihoods, aspirations and level of development (across the globe and within nations) has consistently posed complications in the promotion of a uniform climate change adaptation momentum (Casado-Asensio and Steurer, 2014; Sweeney *et al.* 2013). The most powerful groups in the society who have done well from existing arrangements are very cautious about disturbing the status quo (Sweeney *et al.* 2013), for example, the United States of America and China were very conscious in signing the Paris Agreement which came into on 4 November 2016 after the feet-dragging by the two big carbon emitters (). Climate change governance requires governments to take an active role in bringing about shifts in interest or perceptions so that societies can support or deploy resources for active mitigation and adaptation (Casado-Asensio and Steurer, 2014).

Actors in Climate Change Adaptation (State and Non-State Actors)

Climate change is a global problem that requires global solutions, but the nature of the problem and its impacts require the active involvement of multiple national and local-level stakeholders in shaping and implementing the solutions (Casado-Asensio and Steurer, 2014). The forthcoming figure shows a diagrammatic representation of the actors in climate change.

Acknowledging that, climate change mitigation and adaptation requires a combination of efforts, many players to make this a reality. There are state actors which can be called the government (central and local) (Koch *et al.* 2013). The government has to help in improving climate change knowledge; without government intervention, too little information would be generated for example the information on the need to reduce greenhouse gases is best when supported by the state (*ibid.*). The government also provides the regulatory framework for climate change adaptation. The regulatory and legislative framework act as an incentive or disincentive directing development and human activities in the direction that mitigate climate change (Forino *et al.* 2015). Understanding that the state, in all countries, is very a very big institution for example has many ministries and departments within the ministries, has many agencies and boards meant for various activities. There is need for internal coordination in terms of policies, legislation, efforts and activities of the state (Casado-Asensio and Steurer, 2014). Also, the state has the mandate to relate the national climate change agenda to the regional and global efforts to synchronise the local climate change governance to the transnational governance of climate adaptation (Gukurume, 2013).

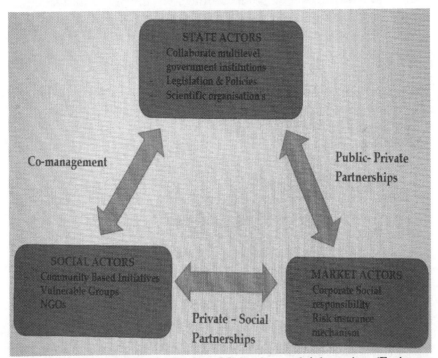

Figure 1: Actors in Climate Change Mitigation and Adaptation (Forino *et al.* 2015: 381).

Apart from the state actors, there are non-state actors in the climate change mitigation and adaptation. Non-state actors represent a range of interests and discourses (Hall, J. and Murphy, 2012a). Their activities take place at different levels ranging from local to global. These non-state actors come in form of market actor and social actors (Forino *et al.* 2015). The market actors include the private entities or corporation that take climate change as a social responsibility. It is appreciated that entities are contributing so much to the climate change through their business operations (*ibid.*). Their compliance to the climate change laws, agreements and policies is one such step, in promoting climate change mitigation for example the cutting of greenhouse gas emissions; industrialists are in the process to change their production methods to be environmentally friendly (Koch *et al.* 2007). Also, new inventions that favour climate change mitigation for example the increased manufacture of solar powered gadgets, new machines (energy efficient) and technologies with the effect of reducing the carbon footprint are other efforts by the market actors (*ibid.*). The private entities have invested in risk and insurance mechanisms which have helped in climate change adaptation. Insurance policies have increased the capacity of many individuals to recover from losses emanating from climate change catastrophes.

Still as non-state actors, there are those which can be termed social actors. These are mainly composed of the Non-Governmental Organisations, Community-based Organisations and other charity organisations who have voluntarily taken the initiatives of climate change adaptation. Globalization processes have rendered non-state actors an integral part of global governance. These "political entrepreneurs" are important players that carry out diverse roles, including information sharing; capacity building and implementation; and rule setting (Andonova *et al.* 2009). Non-state actors, therefore, play diverse roles across the whole policy spectrum from influencing policy-makers to taking action independent of states. This part of non-state actors is instrumental in shaping the contours of climate governance, including through community-based governance (*ibid.*). Community-based Organisations have been instrumental in equipping the local people with competencies and capacities to mitigate and adapt to climate change (Zigomo, 2012). This managed to tap the community-based initiatives towards climate change adaptation for example in Zimbabwe, in farming; NGOs have promoted conservation farming method called *makomba*. This method makes the dig (till), apply manure and fertiliser only were the seed is to be planted only. This has increased yields thereby protecting people from food shortages emanating from the recurrent droughts (*ibid.*). Also, the NGOs have assisted the vulnerable groups with aid, improving their capacity to adapt to climate change.

Coordination as a Key Factor in Bringing Together Stakeholders
National governments, in conjunction with local government, NGOs, the business world and other non-state players, are required

to play their part in the fight against climate change and adaptation to the same; this points to the need for coordination of efforts (Andonova *et al.* 2009). Also, in as far as climate change mitigation and climate adaptation measures are necessary, they require coordination with other spheres of activity and land uses in order to resolve or minimise potential conflicts (Hall and Murphy, 2012a; Sweeney *et al.* 2013). Learning that complex environmental challenges cut horizontally across sectors and vertically across levels of government, to address them, there is need for coordinated and integrated approaches (Casado-Asensio and Steurer, 2014). The plans and policies to tackle environmental problems such as pollution, waste management and protection of environmental sensitive areas, among others, had traditionally been command-and-control regulations that usually prescribed end-of-pipe measures (*ibid.*). However, there are complex environmental problems emerging, such as biodiversity loss or climate change, which affect several policy sectors and all levels of governance. This therefore requires a change in the patterns of governance, policymaking, economic practices, social norms and individual behaviours (Delmas and Young 2009); production of comprehensive responses matching the spatial and sectorial scopes of underlying problems.

Climate change mitigation aims at limiting global warming by reducing global greenhouse gas emissions and enhancing sinks (Wilbanks and Sathaye, 2007). Since greenhouse gas emissions result from the activities of many sectors, integrated mitigation strategies ideally prioritise action in high-emitting sectors, such as energy, transport and industry (*ibid.*). Finally, adaptation refers to the adjustments of natural and human systems in response to climate change (Birkmann and Mechler, 2015). Since adaptation requires more variegated and context-related responses than those taken in sustainable development or mitigation, integrated adaptation strategies are ideal typically "top-down instruments that frame bottom-up measures" of adaptation (Sweeney *et al.* 2014). The adaptation policies previously were designed through top-down and bureaucratic models. Consequently, adaptation policies appeared as an imposition of state views on private affairs and an unnecessary administrative burden to many actors in the political system, even to the vulnerable actors who are supposed to benefit from them (Wilbanks and Sathaye, 2007; Sweeney *et al.* 2014). At this stage, it would certainly lessen the scepticism and opposition toward adaptation if these policies would focus more on inciting and convincing individuals to autonomously integrate climate impacts prevision in their everyday business (Sweeney *et al.* 2014). This can be done through bringing together all the stakeholders, stakeholder participation in formulating comprehensive environmental plans and policies.

Physical Planning as Idea Translation Medium

Physical or spatial planning influences the spatial distribution of activities; creates a more rational territorial organisation of land uses and the linkages between them, to balance demands for development with the need to protect the environment (Economic Commission for Europe, 2008). Taking from its roles, physical planning enhances the integration between sectors such as housing, transport, energy and industry and to improve national and local systems of urban and rural development paying attention to environmental considerations. Spatial planning is therefore an important lever for promoting sustainable development and improving the quality of life hence can be used as a conduit through which the environmental plans or ideas into reality (Economic Commission for Europe, 2008; Chirisa *et al.* 2016). Measures to help effect such a change include regional plans, master plans, local plans, planning schemes and by-laws among other instruments; termed development control mechanism (Chirisa *et al.* 2016). Given that the anthropogenic emissions responsible for climate change are resulting from the development which in effect is regulated by spatial planning through development control (IEED, 2013). Physical planning becomes an entry point through which such activities can be regulated for example integrated land use and transport planning reduces the emission from transport.

Spatial planning is a key instrument for establishing long-term, sustainable frameworks for social, territorial and economic development both within and between countries (Rankomise, 2015; Economic Commission for Europe, 2008). The recent physical planning efforts are based on the efficient use of resources, good governance, public-private partnerships and effective decision-making with regard to investments. In translating the climate change mitigation and adaptation, physical planning has a regulatory and a development function (Sweeney *et al.* 2013). As a regulatory mechanism, government (at local, regional and/or national levels) has to give approval for given activity (in line with the vision of environmental protection and climate change mitigation) (Mkandla,

128

2014). As a development mechanism, government has to elaborate upon development tools for providing services and infrastructure, for establishing directions for urban development, for preserving national resources and for establishing incentives for investment in climate change mitigation measures (*ibid.*). One function of planning is coordination; effective spatial planning aids in the prevention of duplication of efforts by actors such as government departments, commercial developers, communities and individuals (Chirisa *et al.* 2016; Koch *et al.* 2007; Economic Commission for Europe, 2008). This is of great importance, as climate change issue are of a cross-sectorial nature and therefore should be treated as such. Physical planning has the potential, in an integrated and coordinated matter, to prepare regulatory planning instruments and establish priorities for action. It can also facilitate the preparation of local spatial plans, coordinate planning with neighbourhood authorities and engage with the community using participatory planning techniques. Moreover, it can take proactive measures to encourage development and monitor the implementation of policies and proposals by enforcing adherence to specific planning legislation. All these work as a positive move in climate change mitigation and adaptation (Economic Commission for Europe, 2008).

Methodology

The study is a qualitative enquiry which employed a case method and narratology. Case study research excels at bringing a deeper understanding of a complex issue or object through analysis of the experiences or what is already known through previous researches in Zimbabwe. This formed an empirical inquiry that investigated the climate change phenomenon within its real-life context employing multiple sources of evidence. This research approach emphasized detailed contextual analysis of climate change mitigation and adaptation events or conditions and their relationships. The institutions gave a rich content on the current experiences in the complex phenomena of climate change adaptation and mitigation and also aired their views on the relevance of physical planning being the coordinating arm in the climate change mitigation efforts. The data sought was cleaned and collated to be presented in themes. Analysis was done using the discourse analysis.

Results and Discussion

Zimbabwe is a vibrant player at various levels in the international and regional environmental arena (Mkandla, 2014). It acts in various levels including technical level, the sector inter-state policy level (in ministerial forums such as the African Ministerial Conference on Environment [AMCEN], the African Ministers' Council on Water [AMCOW] and others dealing with energy, agriculture, housing and urbanisation, disaster reduction, meteorology among others) and at the continental and global summits (*ibid.*). At the regional level Zimbabwe is party to processes, instruments and agreements in the context of the Southern Africa Development Community (SADC) and the Common Market for Eastern and Southern Africa (COMESA). Locally, it has enacted several laws and bye-laws and set up institution for climate change response for example the Climate Change.

Stakeholders Involved in Adaptation Governance in Zimbabwe

Climate variability and change pose a significant threat to sustainable development and poverty reduction in Zimbabwe. In this vein, a wide range of stakeholders have recognised the threat posed by climate change and have begun to respond at least by addressing agricultural vulnerability, reduction of anthropogenic gases and climate change adaptation information dissemination (Rankomise, 2015). There are state and non-state stakeholders in climate change adaptation. The government and its subsidiaries form the state actors whilst the non-state actors include all individuals and organisation outside the institution of government who either voluntarily or as a business enters in the fight against climate change (Rankomise, 2015; Gukurume, 2013). The relationship that exists between these players is yet to be fluid; the government is mostly regulatory figure in the whole framework and sometimes financing (Gukurume, 2013).

Table 7.1: State Actors in Climate Change Mitigation and Adaptation

Area	Actor	Responsibility
Agriculture (Food Security issues)	Ministry of Agriculture, Farm Mechanisation and Irrigation Development	- Maintaining the strategic grain reserve. - Promoting sustainable farming
	Ministry of Lands and Rural Resettlement	- acquire, distribute and manage the agricultural land resource
	Ministry of Local Government	- Planning for emergencies and disasters.
	Ministry of Environment, Water and Climate	- Regulate utilization and management of water resources.
	Ministry of Energy	- Supplies of electricity
	Ministry of Transport and Infrastructural Development (Meteorological Service Department)	- Provision of meteorological, climatological and seismological services
	Ministry of Finance	- Financing agriculture and food aid programmes.
	Ministry of Media, Information and Broadcasting Services	- Climate change information dissemination
Atmosphere (Greenhouse gases emission reduction)	Ministry of Environment, Water and Climate	- Regulation of GHG emission levels.
	Ministry of Local Government	- sustainable land-use planning
	Ministry of Energy and Power Development	- supply of clean energy sources
	Ministry of Transport and Infrastructural Development	- provision of sustainable transportation services
Biodiversity (Carbon sinks protection)	Ministry of Environment, Water and Climate	- Protecting biodiversity/forests
	Ministry of Lands and Rural Resettlement	- Responsible distribution of land resources
	Ministry of Local Government	- Sustainable land-use planning
	Ministry of Energy	- Supplies of electricity
Climate Change Disaster Response	Ministry of Local Government (DPP and Civil Protection Unit)	- Planning for emergencies and disasters.
	Ministry of Finance	- Financing disaster response efforts
	Ministry of Defence	- Providing manpower in

131

		catastrophes
	Ministry of Transport and Infrastructural Development (Meteorological Service Department)	- Provision of meteorological, climatological and seismological services
Water	Ministry of Environment, Water and Climate	- Protection of water sources and provision of bulk water.
	Ministry of Local Government (DPP and Civil Protection Unit)	- Planning for water supply infrastructure. - Planning for emergencies and disasters.
	Ministry of Finance	- Financing water infrastructural development.

NB: *these actors are by no means exhaustive but a reflection.*

From the table above, it can be appreciated that the state is acting through its ministries and departments of the ministries. Most of the roles on mitigation and adaptation area are stationed in the Ministry of Environment Water and Climate (MoEWC). This ministry has a mission to develop, implement and coordinate policies pertaining to environment, water and climate, to create a clean, safe and healthy environment (MoEWC, 2016). It is also into the management, conservation and the sustainable use of natural resources. This is the arm of government that domesticate multilateral and regional protocols and agreements that Zimbabwe has ratified (*ibid.*). Mitigation is done by the legislation of the Environmental Management Act. This ministry, among other issues, regulates the GHG emissions by industrialists and development agents. The ministry of energy and power development has also a mandate to continuously supply electricity and find cleaner sources with the impact of reducing gaseous emissions which are detrimental to the environment for example the mooted Solar Power Station in Gwanda, expansion the Kariba Hydro-electric power plant and photovoltaic power plant constructed in Marondera by Green Rhino Energy (*Herald*, 21 January 2016). Less spoken is the role by the Ministry of local government, Public works and National housing which houses the Department of Physical Planning (DPP). This ministry, through DPP, has the responsibility to plan for sustainable settlement development hence has the potential to institute designs that reduce gaseous emissions for example compact development that limits the use of automobiles in transportation.

In terms of adaptation, the other ministries also take part including the Ministry of Agriculture, Farm Mechanisation and Irrigation Development (MoAFMID). The ministry has embarked on a number of projects to sensitize farmers on drought resistant varieties, construction of irrigation schemes and provision of farm machines (MoAFMID, 2016). On top of this, it is charged with the maintenance of the country's grain reserves and provision of the food aid in times of climate change induced food shortages. The ministry of finance has also been involved in financing food acquisition efforts and its distribution for example in 2016 it set aside around US$1 billion, under the coordination of the Reserve Bank to ensure food security at both household and national level (*ibid.*). In addition, it funds the disaster response initiatives by the other arms of government for example the floods in Tokwe-Mukosi it mobilised about mobilising $12 million for the cause (*Newsday*, 21 January 2016). The Ministry of Transport and Infrastructural Development (Meteorological Service Department) through the Meteorological Services Department also assists in the disaster preparedness with which the mandate is stationed with the Ministry of Local government, Public Works and National Housing, the Civil Protection Unit. Civil Protection Unit (CPU) coordinates and promotes strategic planning for emergencies at the individual, community, sectorial, local authority and national levels through regulatory mechanisms in order to provide for and ensure optimal emergency preparedness and disaster prevention in Zimbabwe (CPU, 2016). This is from a belief that the Government of Zimbabwe has the primary role to saving lives, protection of property and the preservation of the environment.

Climate change is also prompting a growing number of non-state actors including NGOs and research organisations, including UN agencies, to build strong adaptation components into new development projects, particularly in drought-prone rural areas (Rankomise, 2015; Gukurume, 2013). The non-state actors are more on communities' capacity building for them to effectively respond tom climate change impacts or mitigate its impacts. They spearhead community-based adaptation to climate change (Gukurume, 2013). This approach recognises that climate change impacts will fall hardest on those who are least able to cope. Hence responses will require local adaptation planning and a greater focus on building adaptive capacity and that individuals and communities already have a strong reservoir of skills and knowledge that could increase their resilience (IEED, 2013; Mkandla, 2014; Rankomise, 2015). The NGOs, with their good financial accounts, are the key players in equipping most rural areas in climate change adaptation. They can do so by educating the population of Zimbabwe about climate change effects, mitigation and adaptation. For example, the Coping with Drought and Climate Change in Chiredzi District project demonstrates how community-based adaptation can empower local farmers, ensuring they actively participate in developing culturally sensitive and locally appropriate adaptation strategies for future climatic changes (Gukurume, 2013).

Local Authorities of major urban and mining cities such as Harare, Bulawayo, Gweru, Mutare, Hwange and Kwekwe, and Local Authorities, like the Ministry of Health and Child Welfare, collect data emissions (both dust and gases) as it relates to human health. The Urban Councils Act and the Rural District Councils Act empowers local authorities to make bye-laws relating to the management and conservation of indigenous resources (IEED, 2013; Rankomise, 2015). This provides them with a legal framework to take part in the mitigation of climate change through bye-laws that monitors and keep emission levels at levels that do not disturb the environment. The local authorities are also charged with the planning of the settlement within their jurisdiction which makes them an important player in climate change mitigation and adaptation (Gukurume, 2013). Overall, the regulatory framework in climate change governance has been top-down and usually dominated by the state (*ibid.*). Civil society organisations and local communities have so far played a limited role in the formulation of national climate change adaptation policies and strategies. This situation undermines key governance principles such as equity, stakeholder participation, accountability and transparency (Mkandla, 2014). Stakeholder needs and interests are therefore not adequately reflected in adaptation responses (*ibid.*).

Coordination in Climate Adaptation in Zimbabwe

The Ministry of Environment, Water and Climate is mandated to develop, implement and coordinate policies pertaining to environment, water, climate, creating and maintaining a clean, safe and healthy environment, ensuring management, conservation and the sustainable use of natural resources (MoEWC, 2016). It is the

134

custodian and coordinator in environmental issues and climate. This ministry houses the department which is specifically meant for climate change named Climate Change Management (*ibid.*). Given that climate change adaptation and mitigation issues are cross cutting, there is no comprehensive policy that coordinates the efforts of the players involved. Some of the departments or activities directly involved in climate change mitigation are stationed in different government ministries. For example, the Civil Protection Unit which responds to climate change disasters like floods or cyclones is in the Ministry of Local Government, Public Works and National Housing. As well, the Meteorological Services Department, which also help in the preparedness and consciousness in climate change, is housed in the Ministry of Transport and Infrastructure Development (Gukurume, 2013). This creates operational inefficiencies. Above all, however, it is noted that the MEWC is the oversea and coordinator accountable of all environmental issues and climate change. It regulates all the activities that seem to affect the environment (*ibid.*).

Not underestimating the coordination and complementarity among the players in the fight against climate change, there are bottlenecks in the system including politicisation of the phenomena, too much talk and no action, corruption, financial constraints and jurisdictional overlaps, duplication and conflicts (Rankomise, 2015). It is on most cases that the lack of a comprehensive policy, brought about by the co-effort of all the stakeholders that derails adaptation and mitigation efforts in Zimbabwe. The state, on regulation, it is usually too imposing with little consultation and nurturing of community-based initiatives in climate change adaptation and mitigation (Rankomise, 2015; Gukurume, 2013). There are several concerns raised on transparency and accountability by the state in the enforcement of the laws and distribution food aid. It has, on many cases, the politicisation and corruption has affected a holistic promotion of climate change adaptation. Taking for example in the climate change mitigation, the authorities involved are blamed of being selective in application which has been attributed to corrupt tendencies within the system of government (*Daily News*, 8 September 2016). The food aid, also, it has been said it is being distributed along political lines (*ibid.*). These allegations have actually reduced confidence in the system of state institutions in assisting climate change victims. The government being compounded with liquidity constraints, again, this has affected its efforts in the fight against climate change. This has made the non-state players, especially the NGOs and international organisations being the major players in improving the capacity of locals to adapt to the ravages of climate change (Zigomo, 2012).

The Role of Physical Planning and Civil Protection in Zimbabwe

The Department of Physical Planning and Civil Protection Unit are stationed in the Ministry of Local Government, Public Works and National Housing. The Department of Physical Planning (DPP) is in charge of the spatial planning in the country. DPP influences the spatial distribution of activities in cities, urban areas and rural local authorities through the master plans, town planning schemes and local development plans. DPP has planners who are in charge of planning on state land and provides a monitoring function on the local planning authorities (Chirisa *et al.* 2016). The influence of DPP on climate change is phenomenal, though still not recognised in the mainstream climate change mitigation or adaptation efforts (*ibid.*). The physical planning as an idea translation medium by its abilities to shape the built environment present much room for it to create sustainable settlements which are climate change resilient and with less carbon footprint (Gogo,2 September 2013). At the moment, due to a lack of political will and a comprehensive policy framework, the DPP has not been championing creation of settlements that have reduced carbon footprint as witnessed by sprawl development which is known for being automobile dependent (*ibid.*). Transport emissions forms a greater part of the GHG emitter in the country. The DPP has the potential of creating an environmentally friendly settlement with land uses distributed to promote climate change resilience. For example, preservation of wetlands, promotion of compact settlements that limit transport related emissions and creation of vibrant marketplaces that economically empowers the communities to recover from climate change catastrophes (Chirisa, 2014). High urban poverty is partly antecedent of the distribution of the urban land uses. For example, transport poverty has contributed immensely to the deprivation of access by many to services that allow them to be resilient to climate change induced problems like food shortages (*ibid.*).

The Civil Protection Unit is mandated to discharge overall coordination of all stakeholders involved in disaster risk

136

management, promote preparedness planning, prompt emergency response, early recovery and rehabilitation of affected elements and advocate for integration of disaster risk reduction into development for sustainability (Rankomise, 2015; CPU, 2016). This unit is very important in disaster preparedness and planning for response with the aim of an early recovery for example during floods. It has a close working relationship with the Meteorological Services Department for weather forecast which assist in preparing for disasters including mobilisation of resources and awareness. According to the Civil Protection Unit (2016) its mandate is too swiftly and effectively rescue to victims' disasters (climate change included) and to strengthen coping capacities of the general public in relation to the country's risk profile. In addition, it is also there to design early warning mechanism at all levels, to promote indigenous knowledge systems and documentation of same for posterity and to ensure communities are equipped with basic knowledge and skills to manage prevailing hazards. Besides, it has also a role to develop preparedness plans at national, provincial, district, local authority, community and strategic institutional levels. The primary aim being to save lives, protection of property and the preservation of the environment (*ibid.*). Due to underfunding and poor forecasting in events like floods and cyclones, the responses have not been effective and recommendable for example the flood victims of Tokwe-Mukosi in 2014 are yet to recover from such an event (Maponga,23 March 2016). This is a sign of compounded response or adaptation to climate change risks.

Conclusion, Policy Options and Practical Recommendations

The study exposed that proper societal transformation is a production of committed coordination of which meaningful climate change adaptation and mitigation is the foundation. In Zimbabwe, coordination is not a given and most of the initiatives in climate change adaptation, having a physical planning import, are often overlooked. The harmonisation and consequent fine-tuning of sector policies and strategies will no doubt facilitate identification of synergies between and among sectors in terms of disaster preparedness and reducing the impact of climate-change induced disasters. This is important because challenges identified in some sectors may find ready solutions in others.

The crosscutting impacts of climate change and the imperative need for an integrated response requires resilient and adaptive institutions and exemplary actors to lead the process towards creating an enabling environment for adaptation to climate change. Inadequate institutional support and inappropriate policies can act as a constraint to adaptation and limit access to much needed natural resources by communities dependent on such resources for both survival and adaptation to environmental change and climate variability. It is therefore recommended that a clear and comprehensive policy framework which is adaptive (not rigid) be put in place to tackle climate change issues. This policy framework would bring all the state actors involved in climate change adaptation under one roof for easy coordination of efforts. The policy framework must emphasise participatory approaches that has the impact of building on local knowledge, promote mutual learning and also promotes highest level of transparency and accountability.

Upon observation that, positioning of climate change adaptation within the environment sector limits effective integration. The placement of the climate change adaptation solely within the environment sector with little reference to other sectorial plans is problematic. This limits public and decision-makers' understanding of climate change impacts and the implications for national economies. This undermines political buy-in for prioritisation and resource mobilisation for climate change adaptation. Often guidelines for mainstreaming climate change adaptation into national level planning are not availed to economic planners. Addressing the impacts of climate change and planning for adaptation is therefore done *ex post facto* and in an ad hoc manner. It is therefore recommended that climate change be placed not in the line ministries but in the office of the president and cabinet for effective integration in all parts of the government. It being placed in the current ministries has left many powerful political figures, influential businesspeople and general populace disregarding efforts in this environmental management. It is also recommended that physical planning be empowered to be key player in climate change mitigation. The influence of spatial planning on the built environment is tremendous hence its influence towards climate change resilient settlement cannot be underestimated.

References

Andonova, l. B., Betsill, M. M., and Bulkeley, H. (2009). Transnational Climate Governance. *Global Environmental Politics*, 9(2), 52 73.

Birkmann, J., and Mechler, R. (2015). Advancing climate adaptation and risk management. New insights, concepts and approaches: What have we learned from the SREX and the AR5 processes? *Climatic Change*, 133(1), 1–6.

Casado-Asensio, J., and Steurer, R. (2014). Integrated strategies on sustainable development, climate change mitigation and adaptation in Western Europe: communication rather than coordination. *Journal of Public Policy*, 34 (3), 437-473.

Chirisa, I., Bandauko, E., Mazhindu, E., Kwangwama, N. A., and Chikowore, G. (2016). Building resilient infrastructure in the face of climate change in African cities: Scope, potentiality and challenges. *Development Southern Africa*, 33 (1), 113-127.

Civil Protection Unit. (2016). Available online: www.zimdrm.gov.zw/. [Accessed 29/11/16] .

Daily News. (8 September 2016). *Zanu-PF politicising food aid*. Harare: Alpha Media Holdings.

Delmas, M., and Young, O. (2009). *Governance for the Environment: New Perspectives*. Cambridgeshire: Cambridge University Press.

Economic Commission for Europe. (2008). *Spatial Planning Key Instrument for Development and Effective Governance with Special Reference to Countries in Transition*. New York and Geneva: United Nations.

Forino, G., von Meding, J., and Brewer, G. J. (2015). A Conceptual Governance Framework for Climate Change Adaptation and Disaster Risk Reduction Integration. *International Journal of Disaster Risk Sci*, 3(6), 372–384.

Giddens, A. (2008). *The Politics of climate change – policy network*. Available online: http//:www.policynetwork.net/.../The_politics_of_climate_change_Anthony_Giddens.pdf. [Accessed on 20/11/16]

Gukurume, S. (2013). Climate change, variability and sustainable agriculture in Zimbabwe's rural communities. *Russian Journal of Agricultural and Socio-Economic Sciences*, 2(14), 89-100.

Hall, J., and Murphy, C. (2012a). *Adapting Water Supply Systems in a Changing Climate. In: Water Supply Systems, Distribution and Environmental Effects*. New York: Nova Science Publishers.

IEED. (2013). Climate change responses in Zimbabwe: local actions and national policy. Available online: http://pubs.iied.org/17145IIED. [Accessed on 28/11/16]

Koch, I.C., Vogel, C., and Patel, Z. (2007). Institutional dynamics and climate change adaptation in South Africa. *Mitigation and Adaptation Strategies for Global Change*, 12 (8), 1323–1339.

Ministry of Environment, Water and Climate (MoEWC). (2016). Mission, Functions and Roles. Available online: http://www.environment.gov.zw/. [Accessed on 29/11/16]

Mkandla, S. (2014). Climate change: Aligning policy and residual knowledge with practice in Zimbabwe. Paper presented to the conference themed Disaster Reduction and Emergencies – Preparing for climate change eventualities held on 23 October 2014 in Harare. Available online: bulawayo24.com/index-id-opinion-sc-columnist-byo-56368.html. [Accessed on 28/11/16]

Newsday. (21 January 2016). *ZPC Finalising funding for Gwanda Solar Plant.* Harare: Alpha Media Holdings.

O'Brien, G., O'Keefe, P., Rose, J., and Wisner, B. (2006). Climate change and disaster management. *Disasters*, 30(1), 64-80.

Rankomise, A. O. (2015). Climate Change in Zimbabwe: Information and Adaption. Available online: www.kas.de/Zimbabwe/. [Accessed on 28/11/16]

Sweeney, J., Bourke, D., Coll, J., Flood, S., Gormally, M., Hall, J., McGloughlin, J., Murphy, C., Salmon, N., Skeffington, M. S., and Smyth, D. (2013). *Co-ordination, Communication and Adaptation for Climate Change in Ireland: An Integrated Approach.* Wexford: Environmental Protection Agency.

Tanner, T., and Allouche, J. (2011). *Towards a New Political Economy of Climate Change and Development.* New Jersey: Blackwell Publishing Ltd.

Gogo, J. (2 September 2013). Urban Areas must plan for climate change. *The Herald*, Harare: Zimpapers.

Maponga, G. (23 March 2016). *Lifeline for Tokwe-Mukosi Flood Victims. The Herald*, Harare: Zimpapers.

Wilbanks, T. J., and Sathaye, J. (2007). Integrating mitigation and adaptation as responses to climate change: a synthesis. *Mitigation and Adaptation Strategies for Global Change*, 12 (5), 957–962.

Zigomo, K. (2012). A Community-Based Approach to Sustainable Development: The Role of Civil. Available online: http://www.solidaritypeacetrust.org/1159/community-based-approach-tosustainable-development/. [Accessed 29/11/16]

Chapter 8

Urban Planning Tools for Climate Risk Management in Zimbabwe

Sharon Marimira, Chipo Mutonhodza & Thomas Karakadzai

Introduction

Cities are centres for social, economic and cultural development but climate risks are challenging their role as growth engines and innovation hubs. Urban areas suffer a lot from the effects of climate change due to their concentration of infrastructure and assets, aggregation of people and continuous urban expansion. And this has been made worse by the fact that more than half of the world's population lives in cities, with an additional 2 billion urban residents expected in the next 20 years World Bank (2009). Average densities for urban areas in Africa are expected to increase from 34 to 79 persons per square kilometre between 2010 and 2050. 78% of Europeans live in urban areas and nearly 85% of the European Union' total GDP is being generated in cities. This makes urban vulnerability to climate change a major challenge for all cities (Eurostat, 2012; ADB, 2012).

Urban areas are significant producers of GHSs and making them hotspots for climate disaster risks such as natural disasters and extreme weather conditions such as floods, droughts and famines, pestilence and major storms, which have resulted in economic, social and physical losses, destroying development gains and exacerbating poverty. Kreft *et al.* (2016) argue that of the ten often most affected countries by climate change extreme weather events between 1994 and 2013 nine are developing countries with high numbers of low-income groups, like Honduras, Myanmar, Haiti, Philippines, Cambodia and India resulting in more than 530,000 people dying and costing about 2.2 trillion USD in damages.

The challenge with urban areas is that the majority of them have high level of informal settlements which increase exposure to climate risk as these are areas of high population density. Dickson *et al.* (2010) state that about a billion people live in informal settlements which areas lack basic, essential infrastructure and services. For example, 28 percent of the 15 million of Dhaka live in informal settlements, in South Africa 250 000 to 300 000 families live in informal settlements, Habtezion and Padgham (2014). The problem is that the settlements are located in high-risk areas and are constructed in faulty shelters, which have limited access to basic and emergency services, Dickson *et al.* (2012). The slums are situated in marginal areas which are steep hillsides, flood plains, near hazardous waste and coastal zones which are at great risk from flash floods, landslides and heavy downpours which are impacts of climate change, Dickson *et al.* (2012) and Baker (2012).

Cities in developing countries struggle to meet basic needs of their residents, such as clean water, sanitation and mass transport and with the increase in climate change impacts there would be an increase in these urban challenges. These present themselves via indirect impacts such as blackouts, increased risk of water-borne diseases, heat stress and destruction to transport infrastructure affecting travel to work and preventing goods to reach the market, ADB (2014). While an increasing number of cities are prioritizing sustainable low carbon development and signatories to Resilient Cities Campaign by the United Nations International Strategy for Disaster Reduction (UNISDR) many do not understand the climate risks they are exposed to and lack the necessary adaptation and resilience policies (Mabey *et al.* 2014). Therefore, this chapter seeks to analyse how urban planning tools in Zimbabwe manage or adapt to climate change so as to pave way to creating climate change resilient cities. Land use planning can be used to enable both local and national adaptation to climate change though official plans, zoning and development permits to minimise risks to communities from floods, wildfires, landslides and other natural hazards. This chapter is based on research done using literature and document review, which interrogated existing published and unpublished documents and databases such as newspapers and journals. Document review was particularly useful in providing estimates of relevant parameters such as current data about what is happening on the ground in terms of climate change adaptation in Zimbabwe.

In terms of organisation, the first part of the chapter introduces the study and the scope of the research in terms of the study aim and objectives and the problem statement. The second part reviews the literature as well as concepts that have been promulgated in relation to the study. The third part focuses on the research methods to be used to conduct the research and the fourth part will present and analyse the data gathered. The last part will analyse the findings of the study and give conclusion and recommendations.

Context of Study

Climate risk refers to threats to human and natural systems resulting from climate change impacts and the vulnerability of these systems in the face of these threats. The risks manifest themselves in direct physical impacts of severe weather events and other climate impacts, as well as second and third order consequences of climate-related events (Mabey *et al.* 2016). These include droughts, veld fires, more frequent cloudbursts and extremely cold winters while other areas experience increased levels of heat waves which impacts public health and transport and power infrastructure.

Because of the densely built environments of cities they are more vulnerable to climate change events, especially to heat waves, as they absorb more heat by concrete and other building materials and worsened by the removal of vegetation and loss of permeable surfaces, creating urban heat islands effect (Rosenzweig *et al.* 2011; Dickson *et al.* 2012; Mabel *et al.* 2016). In Europe, according to the European Commission, at least 1,000 km² of land converted for new infrastructure every year. In Mbabane City, the city boundary in terms of coverage in square kilometres increased from 39.07 square kilometres in 1976 to 63.87square kilometres in 2006. Increased surfaced areas increase the severity of floods and they also limit the soil drainage capacity. Therefore, this reduces the amount of the underground water level and the high concentration of population in urban areas (high density) worsens water scarcity due to increased demand.

Another climate risk faced by urban areas according to Mabel *et al.* (2016) is that cities form highly interdependent and interconnected systems like sewage systems, electricity grids, transportation networks, communications infrastructure and water supply and a failure in one system can lead to cascading effects through a city's entire organism. For example, torrential rains that hit Beitbridge in 2013, damaged, power lines, roads, sewer pipes and bridges forcing the closure of the district hospital and the border. In Muenster, Germany, a severe snowfall in November 2005, caused power lines and communication wires to collapse resulting in a blackout cutting off electricity and communication networks for a quarter of a million people for several days (Zukunftsforum Öffentliche Sicherheit, 2008).

Legislative framework

The beginning of the 20th century saw an increase in natural disaster losses, continuing its upward trend in the present century owing to climate change, which has increased the incidence of disasters, especially meteorological ones such as floods (Ejeta *et al.* 2015). On a global scale, Europe has experienced a 60% increase in extreme weather events causing damages of over €14 billion in UK alone and a heat wave that hit Europe in the summer of 2003 resulted in between 55,000 and 70,000 deaths. The resulting challenge is that they are occurring in faster sequences such as Copenhagen, which suffered a severe 1 in 100 years flood in 2010 followed by an even greater 1 in 1000 years flood the following year (Jendritzky, 2007; Robine *et al.* 2008; EEA, 2012; EASAC, 2013; Haghighatafshar *et al.* 2014).

Initiatives to address the challenges of climate change include the following:

- the Intergovernmental Panel on Climate Change (IPCC),
- the United Nations Framework Convention on Climate Change,
- the Nairobi Work Programme and,
- the Kyoto Protocol and the Bali Action Plan

At the regional level, they include:

- the African Ministerial Conference on Environment (AMCEN)
- the Framework of Southern and Northern Africa Climate Change Programmes and
- the East African Community Climate Change Policy, AU (2014).

The UNFCCC objective is to stabilize GHG concentrations in the atmosphere to a level that would not cause dangerous climate changes and at a timeframe which allows ecosystems to adapt naturally to climate change, food production not be affected and enable economic development to progress in a sustainable manner. While the Kyoto Protocol, which aims at reducing global GHGs by

5% provides the rules and operational modalities on how countries would reduce emissions and measure their emission reductions, a package of laid down detailed rules which include emissions trading, Clean Development Mechanism) and land use, land use change and forestry (LULUCF) (EAC, 2011).

Within the Regional level climate change effects can be noted in Southern Africa which has experienced extreme floods and droughts. For example, in 1982-1983, 1986-87 and 1991-92 serious droughts were experienced in Africa especially in Burundi, Kenya, Uganda and Rwanda that caused a decrease in crop and livestock production. In February 2000, cyclone Eline hit Mozambique, South Africa, Botswana and Zimbabwe, which in Mozambique alone killed more than 700 people, displacing more than a million people and destroyed infrastructure worth $1 billion (Gwimbi, 2009). Regional initiatives include those by the East African Community (EAC) which has developed the EAC Climate Change Strategy and Master Plan with the goal to strengthen regional cooperation through responding to climate change as a shared resource by 2031. The objectives of the plan are manifold. These include

- to provide an effective and integrated response to regional climate change adaptation,
- to enhance the mitigation potential of Partner States in the energy, infrastructure, agriculture and forestry sectors
- to foster strong international cooperation to address issues related to climate change including enhancing the negotiating ability of the Partner States in the African Union and other forums including the UNFCCC and
- to mobilise financial and other resources to implement the above (Otiende, 2013).

Whereas SADC addresses climate change through its Climate Change Adaptation Strategy in the Water Sector and SADC REDD+ Programme. Locally, Zimbabwe has signed and ratified the United Nations Framework Convention on Climate Change (UNFCCC) in 1992 and has also consented to the Kyoto Protocol in 2009. Climate change issues are also broadly included in Zimbabwe's National Environmental Policy and Strategies.

Theoretical Perspectives

Resilience is the capacity of a social structure to absorb disturbances while retaining the same basic structure and ways of operating, the capacity for self-organisation and to adapt to stress and change, IPCC (2007). A resilient city is therefore one which is prepared to absorb and recover from any shock or stress while maintaining its essential functions, structures and identity and adapting and thriving in the face of continual change. Creating resilient cities goes hand in hand with creating sustainable livelihoods, as the framework considers the livelihood context in terms of vulnerability context, assets, policies and institutional strategies designed to address the problems at the local level (Spalivioro et al. 2011; Knutsson, 2006).

Climate change adaptation and management should be initiated in city planning though (1) the formation of city development strategies (2) preparing specific sectorial plans such as transportation, water supply and sanitation, or (3) as a standalone activity integrated into the city's' development strategies or master plans, Dickson et al. (2012). According to Wilson and Piper (2010) land use planning is one of the most effective processes to facilitate local adaptation and management to climate change through enhancing prevention and preparedness and facilitating response and recovery within the community. Land use planning reduces the future carbon impact of new developments as well as for improving resilience against natural hazards associated with climate change and therefore, the link between climate change and land use planning is mitigation while the link between climate change and disaster management is adaptation as shown in figure 8.1, thereby creating a continuum of Prevention, Preparedness, Response and Recovery (PPRR), Bajracharya et al. (2011).

According to World Bank (2012) land use planning aims to: i) identify and mitigate the root cause of disaster risks embedded in existing land development practices through regulated use of land in hazard-prone areas and building codes, ii) reduce losses by facilitating faster response through provision of open spaces and well planned road network for rescue operations and iii) promote controlled urban growth without generating new risks, 'building back better' through rebuilding and upgrading infrastructure using hazard-resistant construction in accordance with a comprehensive plan. Local governments can use land use planning tools (discussed below) such as official plans, zoning, development permits and others to minimise risks to urban areas from floods, wildfires, landslides and other natural hazards.

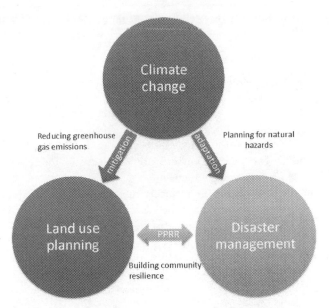

Figure 8.1: The link between climate change, land use planning and disaster management Mitigation (Bajracharya *et al.* 2011)

1. Zoning

Zoning codes can be used by municipalities to limit new development in hazard-prone areas (for instance in areas subject to wildfires, landslides or coastal erosion) or to prescribe building standards that reduce vulnerability to environmental stress (Platt, 1998). For example, in areas of high flood risk, all development may be prohibited, while in areas where the risk is lower, the ground floor of new buildings and structures may be required to be built above a minimum height to avoid flood damage (Kreutzwiser, 2008). Given that climate change often magnifies existing hazard risk, municipalities could respond by modifying existing zoning restrictions to factor in the greater intensity, frequency or duration of certain hazards as climate changes. For example, in Beaubassin-est the bye-law identifies a sea level rise "protection zone" in which the minimum ground floor elevation of any new building must be at least 1.43 meters above the current 1-in-100-year flood mark. Local bye-laws specify in general terms the matters to be considered in such negotiations, giving planners latitude to address a broad range of issues regarding environmental protection, natural hazards and other concerns, Richardson and Otero (2012).

Box 8.1: The case study below shows zoning of hazard prone areas in Philippines. (Istanbul Metropolitan Municipality - Disaster Coordination Centre, WBI, 2009)

In Metro Manila, Philippines, fault zoning in the City of Muntinlupa included demarcation of danger and no-build zones along 10 km. zone, tax relief, relocation and financial assistance to affected residents. However, land surveys conducted by the Zoning Administration Office would usually be unwelcome, notices to vacate danger zones would fall on deaf ears; informal settlements continued to thrive. The government did not acquire land for relocation nor provide sufficient assistance for resettlement.

In Istanbul, the Metropolitan municipality launched a micro-zonation project in the high-risk south-western part; detailed information on local ground conditions were used to establish appropriate design parameters and building codes for construction.

The Map shows the Hazard Zoning in terms of; distribution of peak ground acceleration and number of heavily damaged building, distribution of vulnerable buildings in Sumer neighbourhood; distribution of earthquake weakness scores through the Zeytinbumu District.

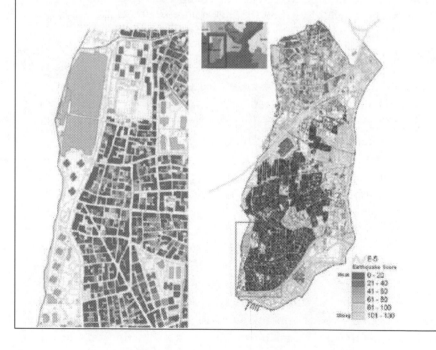

148

Building Codes

Building codes are a critical aspect in adapting to climate change as the International Building Code Council (ICC) argues that building codes provide protection from man-made and natural disasters and keep construction costs at a minimum by establishing uniformity in the construction industry, enabling manufacturers to do business both at a national scale and even international scale. However, in Africa, many of the buildings cannot withstand any natural or manmade disasters if they occur and has resulted in the collapse of many of the buildings. Listokin and Hattis (2005: 21) argue that because of the technical complexity of the codes and the time and money needed to keep them updated most governments in Africa namely Ghana and Nigeria have abandoned the development and maintenance of the codes and resort to just adopting them without amending them. For example, earthquakes occurred in Haiti 2010 killing 222 000 people and injuring 300 572 and in Qinghai, China killing more than 1 700 people (IDRL, 2010). According to CASA (2012) the high level of death in the areas is mainly due to lack of enforcement of seismic building codes that have increased safety through enabling the construction of stronger buildings and infrastructure. According to Listokin and Hattis (2005) the building codes in most countries in Africa, namely Nigeria, Ghana and Zimbabwe in terms of building structural system, fire safety, general safety, enclosure, plumbing code, mechanical codes and electrical codes have not adapted to climate change such that they are vulnerable to climate change weather extremes. For example, according to IDRL (2012) in November 2012 a 5 story Melcom Shopping Centre in Accra, Ghana collapsed killing 14 people and injuring more than 60 people due to lacking enforcement building codes.

Local Plans on Special Matters

These local plans are developed through a formal planning process, including public consultation and provide an assessment of conditions, a long-term vision and a set of goals and actions for the particular issue or area in question. A growing number of Canadian municipalities have adopted plans that specifically address the need to adapt to climate change. Some communities have chosen to develop stand-alone adaptation plans on climate change, Richardson and Otero, 2012).

2. *Covenants and Easements*

Covenants and easements can play an important role in climate change adaptation planning. Covenants may be used in "green developments" that use innovative energy, wastewater treatment and other systems that require public access and the collaboration of future owners for proper operation and maintenance (Titus, 2011; Richardson and Otero, 2012).

3. *Miscellaneous*

Shaw *et al.* (2007) argue that climate change adaptation and management are strongly influenced by urban form as such a sustainable urban setting makes use of water storage capacity, integration of green and blue spaces for cooling and rainfall infiltration. Therefore, there is need for adapting and managing high temperatures, water sources and floods aspects that are climate risks in urban areas. As noted, urban areas are at great risk of the urban heat island effect. Some of the design solutions urban planners can adapt include:

- creating green-spaces, made up of a linked network of open spaces have ecological, recreational and flood storage benefits like open spaces, woodlands, street trees, fields, parks, outdoor sports facilities and community gardens
- creating blue-space, such as open bodies of water, including rivers, lakes and urban canals; shading and orientation to reduce excessive solar gain (e.g. through narrow streets or canopies of street trees) and
- adopting passive ventilation captured through orientation and morphology of buildings and streets (Walsh *et al.* 2003; TCPA, 2004; Shaw *et al.* 2007).

In terms of floods, urban planners can combine these design solutions:

- widening drains to increase drainage capacity;
- managing flood pathways and removing pinch points so that heavy rainfall can drain away and
- encouraging the use of flood resilient materials which can withstand direct contact with floodwaters for some time without significant damage such as concrete, vinyl and ceramic tiles, pressure-treated timber, glass block, metal doors and cabinets.

In managing water resources and quality urban planners can make use of

- controls and licensing to manage the needs of water users while ensuring adequate protection of the environment
- separate drainage systems for surface and foul water so as to send surface water runoff directly back to the watercourse and significantly reduce the treatment burden
- rainwater harvesting and storage from roofs and grey water recycling to use wastewater from plumbing systems for toilet flushing and irrigation (Mostert, 2006; Gill *et al.* 2007; Shaw *et al.* 2007; ODPM, 2004).

The Case of Zimbabwe

Zimbabwe is a landlocked country in Southern Africa, lying between latitudes 15° and 23° south of the Equator and longitudes 25° and 34° east of the Greenwich Meridian. It is bordered by Mozambique to the East, South Africa to the South, Botswana to the West and Zambia to the North and North-west shown in the diagram. It has a sub-tropical climate with four seasons that is cool dry season from mid-May to August; hot dry season from September to mid-November; the main rainy season running from mid-November to mid-March and the post rainy season from mid-March to mid-May. The mean monthly temperature varies from 15° C in July to 24° C in November with annual rainfall averaging at 650 mm (GoZ, 2014, see Figure 1).

According to GOZ (2014), Zimbabwe has experienced a warming trend towards the end of the twentieth century compared to the beginning. Annual mean temperature increased by about 0.4° C, resulting in the increase in both the minimum and maximum temperatures represented by a decrease in the number of days with a minimum temperature of 12° C and a maximum of 30° C, especially during the dry season. And in the same manner there has been a 5 percent decline in rainfall across Zimbabwe as shown in the diagram.

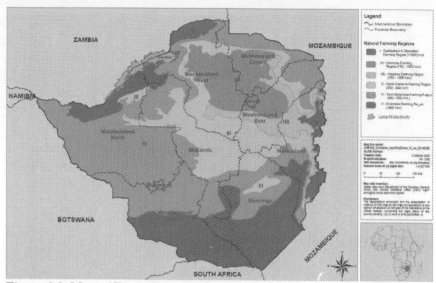

Figure 8.2: Map of Zimbabwe with Ecological regions (GoZ, 2014)

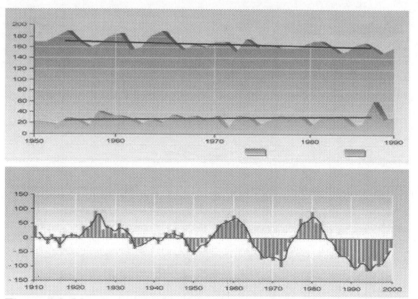

Figure 8.3: Number of days with a minimum temperature of 12° C and a maximum of 30° C during 1950-1990 and national rainfall deviation during 1910-2000 (GoZ, 2014)

Urban areas in Zimbabwe are already feeling the effects of climate and this can be noted through water shortages throughout the majority of them including Harare, Bulawayo, Beitbridge and

152

Mutare. Harare for instance Due to loss of wetlands due to construction of residential areas on their zoned areas has pushed the water table 12 metres deeper to 30 metres. Therefore, in the same manner, it increases the cost of pumping it. These areas are already suffering from poor service delivery in terms of sewerage, solid waste management, water supply and electricity, which are already existing nightmares for cities like Harare. As noted, urban areas suffer a lot from the effects of climate change due to their concentration of infrastructure and assets, aggregation of people and continuous urban expansion city and Harare is one such area that has been expanding horizontally therefore causing serious ecological repercussions and pose a great deal of challenge to environmental sustainability both at a local and global scales. One such effect is that most informal settlements in Harare have a high dependency on natural resources like use of wood fuel, building on wetlands Therefore, destroying land cover and ecosystems which aid in reducing the urban heat island effect and destroying the areas which are sources of their food like agricultural land. Other urban areas like Beitbridge town face heavy rainfalls and floods.

Kamusoko (2013) conducted studies in Harare and documented changes in permanent buildings and the non-built up area for Harare between the years 1984, 2002, 2008 and 2013 and their results show that the built-up area have increased from 118.6 square kilometres in 1984 to 342.2 square kilometres in 2013. The non-built up area has however decreased from 822.9 square kilometres in 1984 to 597 square kilometres in 2013, a change of 23.7% as shown in the Table 8.1.

Table 8.1: Land use and land cover change in Harare (Kamusoko, 2013)

	1984	2002	2008	2013
	Land Coverage km^2	Land Coverage km^2	Land Coverage km^2	Land Coverage km^2
Built Up Area	118.6 km^2	233.9 km^2	302.7 km^2	342.2 km^2
Non-Built Up Area	822.9 km^2	705.6 km^2	636.1km^2	597 km^2

The city boundary of Harare has also been encroaching into surrounding rural areas like Zvimba, Mazowe and Goromonzi. Such actions can be noted by development of residential areas in agricultural land for example Hopley Farm settlement to the south of Harare not only that some areas in Chitungwiza and Hatfield extension have developed residential areas into wetlands. Harare 24 (2012) argues that Harare City Council, in accepting construction projects that have questionable environmental impact assessments on wetlands such as the Millennium Park that will blanket the entire Borrowdale wetland and the Chinese hotel on top of a wetland in Belvedere. The challenge of destroying wetlands is that they are natural water reservoirs that recharge the underground water table, act as carbon sinks, help filter and purify polluted water and absorb excess water when there is flooding.

Odero interviewed in Herald (2013) argues that local cities and towns face uncertain economic and social futures because they are vulnerable to short-, medium- and long-term climate impacts and therefore require seriously re-evaluation of their land use and zoning policies. He argues that cities of the future must be driven by production and consumption patterns tuned to low-carbon sustainable growth. However, Harare is energy inefficient and does not promote conspicuous consumption but rather enables the production of greenhouse gases as not by the horizontal development of the city which creates long distance travel to and from the city. Therefore, more GHGs being emitted into the atmosphere which is a great contributor to climate change.

In addition, such urban forms result in high cost of service delivery shown by the city struggling to provide water and Herald (2013) noted that many townships in Harare go for days and months without running tap water and in 2008 more than 4 000 people died in a cholera outbreak. In addition, the city struggles with effective waste control shown by some parts of the city going weeks without waste collection and landfills which have filled up creating methane gas, Herald (2013). Because of that according to Harare News (2012) more than a million residents of Harare have for the past few years resorted to sinking their own wells and boreholes, an emergency measure that is emptying the underground water table, drying up wetlands and causing environmental damage.as a result, boreholes in Harare are drilled as deep as 100 metres to reach the water table. Such developments have danger for future water supply in that according to the Financial Gazette (2014) a study conducted by the British Geological Survey and the University College London in April 2014, indicates that Zimbabwe's aquifer productivity (litres per second) is so low that it can only

154

produce between 0, 1 and 0, 5 litres of water per second compared to 20 litres per second found in countries such as Angola, Egypt, Chad and Niger.

Another challenge faced by cities in Zimbabwe is the high level of informal settlements that has become the norm of the day. Informal settlements in Zimbabwe were rated at 3.4 percent of the urban population in 2001; 18 percent in 2006 and has risen to about 43 percent (UN-Habitat, 2015). Informal settlements in Harare are in areas such as Dzivarasekwa Extension, Caledonia and Hatcliffe while in Bulawayo include areas like; Cabatsha, Trenance, Ngozi Mine, Durnkirk, Willsgrove and Killarney. Most of the areas are prone to natural hazards both man-made like veld fires and natural like floods as they are located at unsuitable sites such as wetlands and steep slopes. One such example is Ngozi Mine situated near Bulawayo's dump that is near Cowdray Park and is therefore exposed to health hazard from the area. In addition, the areas lack municipal trunk sewers, basic water and sanitation, roads and other forms of municipal infrastructure and services, which all play important roles in mediating disaster risks, especially in hazard-prone areas. Bulawayo 24 (2014) argue that this is attributed to master plans which are now outdated and have failed to effectively regulate development, as demonstrated by the rapid growth of slums and failure for it to facilitate the building-in of resilience against climate-induced disasters and prevent cities from becoming reservoirs of potential epidemics and some areas becoming weak spots in the event of disasters.

Urban areas in Zimbabwe have been responding to climate change through investing in dam construction so as to boost water provision and efficiency and build strong defences against climate change-induced water shortages, already noticed to be affecting crop output and household food security, this can be noted by dam construction and maintenance projects at Tokwe-Mukosi, Osborne, Semwa and Mutange. However, according to Bulawayo 24 (2014) drought conditions created by climate change are expected to reduce run-off, further reducing the water levels required to support the operation of dams. Therefore, this brings the need for Zimbabwe to intensify the mainstreaming of alternative energy sources as a way of hedging against reliance on a restricted energy pool (hydropower) that is vulnerable to climate change. In terms of policy and legislative framework, ZimAsset (2013-2018) upholds the need to establish a national climate change policy to help deal with climate-related disasters but little seems to be done on the ground. According to Herald (01/04/16), the National Climate Change Response Strategy was crafted in 2014. Two years on, it has not been completed.

Planning and design standards used by local authorities are still the same old standards that were used prior independence. The Building By-Laws of Zimbabwe were enacted in 1977 and are still operational to date. The argument in line with climate risks is that they are now outdated to effectively address climate change issues. This can be noted by the Local Development Plan of 1984 used for planning in Beitbridge which is now outdated as there is no hazard zoning of areas prone to flooding and those areas characterized by poor water and sewer reticulation and poor drainage systems (the Mirror, 20/03/16).

According to World Bank (2012) the implementation of land use planning to plan for climate risk crucially depends upon institutional coordination between various sector specific agencies engaged with land management as well as multi-jurisdictional cooperation when risks extend beyond urban limits. Therefore, institutions concerned with land development, including roads and transportation, communication and utilities, housing, should share information and work in a coordinated fashion using an urban information management system that can ensure accountability to residents. The challenge with urban councils in Zimbabwe is that such mechanisms do not exist and what makes the situation worse is that the government ministries like Ministry of Transport, Housing and even Department of Physical Planning view themselves as separate entities from urban councils. Therefore, this makes it difficult to create an integrated comprehensive land use plan to tackle urban climate issues.

Effective climate risk land use planning requires high level institutional capacity, but the challenge faced in urban areas of Zimbabwe is that there is lack of local climate risk information available, making it difficult for them to effectively plan for climate change. Not only that there are limited skilled personnel able to deliver site specific or hazard specific land use plans capable to reduce the risks. In addition to that most land use maps and other relevant risk information like hydrology maps are not up-to-date. According to Madamombe (2007), the early warning system of the Metropolitan Department which is responsible for predicting the weather and flood events has low level of accuracy in terms of forecasting making proactive land use planning difficult.

In addition to that there is lack of financial and technical capability to undertake preventive measures to climate risks. According to Herald (2014) Zimbabwe in the 2015 National Budget has slashed its budget for the environment, water and climate by 44 percent to $52, 7 million from $93 and 5 million in 2013. Therefore, this makes it difficult to build resilience in cities. This has therefore made it difficult for local authorities to secure funds to install appropriate infrastructure and well-advanced technology to be resilient.

Conclusion and Recommendations

This study sought to analyse how urban planning tools in Zimbabwe manage or adapt to climate change so as to pave way to creating climate change resilient cities. Climate change is an inevitable issue that is being faced by urban areas in Zimbabwe. Therefore, this chapter proposes that there is need to create climate risk resilient cities by enabling urban planning tools in Zimbabwe to change and adapt and manage climate risk though prevention, preparedness, response and recovery. The study showed that the planning tools used by urban planners in Zimbabwe are outdated such as their building codes and planning standards and are not effectively addressing the challenges brought about by climate change. The research paper therefore recommends the following;

- Establish zoning codes that limit new development in hazard-prone areas (areas subject to wildfires, landslides)
- Establish building standards that reduce vulnerability to environmental stress, e.g. in areas of high flood risk, all development may be prohibited, while in areas where the risk is lower, the ground floor of new buildings and structures may be required to be built above a minimum height to avoid flood damage
- Establish conservation easements to prevent residential development on wetlands or areas at threat from flooding
- Establishing design guidelines that are climate risk conscious e.g. encouraging the use of building materials that are waterproof or fire resistant

157

- Establish an effective coordinated flow of information between institutions that is government ministries and the local authorities
- Institute Regional Planning Policies and Regulations to which Local and Regional plans must conform. These will address local adaptation efforts for the region like prescribe planning standards designed to protect agricultural land by containing urbanisation at the fridge.
- Enhance the development of Local Plans that adapt to climate change by identifying the impacts of climate change on their local areas and suggesting specific actions to be addressed.

References

African Union (AU). (2014). *African Union Strategy on Climate Change.* AU

Asian Development Bank (ADB). (2014). *Urban Climate Change Resilience: A Synopsis.* ADB, Mandaluyong City

Bajracharya., Childs., and Hastings (2011). Climate change adaptation through land use planning and disaster management: Local government perspectives from Queensland. Paper presented at 17th Pacific Rim Real Estate Society Conference Climate change and property: Its impact now and later 16 -19 January 2011, Gold Coast

Mkandla, S., Alternate Secretary General (ASG)., and ZAPU. (October, 26 2014). *Going Forward: Opportunities and Constraints.* Bulawayo 24, Bulawayo.

CASA. (2012). *Integrating Climate Change and Disaster Risk reduction in Physical development. Review of Ghana Building Code.* CASA, Ghana

Dickson, E., Baker, J. L., Hoornweg, D., and Asmita, T. (2012). *Urban risk assessments: an approach for understanding disaster and climate risk in cities.* The World Bank.

East African Community. (EAC 2011). *Climate Change Strategy.* EAC, Arusha, Tanzania

European Environment Agency. (2012). *Damages from weather and climate-related events (CLIM 039).* European Environment Agency, Brussels.

Ejeta, L. T., Ardalan, A., and Paton, D. (2015). Application of behavioural theories to disaster and emergency health preparedness: A systematic review. *PLoS currents*, *7*.

Government of Zimbabwe (1996). Regional Town and Country Planning Act (Chapter 29: 12), Government of Zimbabwe: Harare

Government of Zimbabwe. (2014). Zimbabwe's National Climate Change Response Strategy. Government of Zimbabwe, Harare

Gill, S., Handley, J., Ennos, R., and Pauleit, S. (2007). Adapting Cities for Climate Change: The Role of the Green Infrastructure. *Built Environment*, 33(1), 115–133

HagHigHatafSHar, S., la Cour Jansen, J., Aspegren, H., Lidström, V., Mattsson, A., and Jönsson, K. (2014). Storm-water management in Malmö and Copenhagen with regard to Climate Change Scenarios. *VATTEN–Journal of Water Management and Research*, 70(1), 159-168.

Kreft, S., Eckstein, D., Junghans, L., Kerestan, C., and Hagen, U. (2016). *Global Climate Risk Index 2016*. German Watch, Berlin

Mabey, C.S., and Schwarzkopf, K. (2014). *Underfunded, Underprepared, Underwater? Cities at Risk*. E3g London

Mostert, E. (2006). Integrated water resources management in the Netherlands: how concepts function. *Journal of Contemporary water research & Education*, *135*(1), 19-27.

ODPM. (2004). *The Planning Response to Climate Change: Advice on Better Practice*. CAG and Oxford Brookes, London.

Wilson, E., and Piper, J. (2010). *Spatial Planning and Climate Change*. Milton Park, Abingdon, Oxon; New York: Routledge. Fidelity Printers and Refiners, Harare

Gwimbi, P. (2009). *Linking Rural Community Livelihoods to Resilience Building in Flood Risk Reduction in Zimbabwe*. Journal of Disaster risk studies, Volume 2 No 1.

Harare 24. (November 2, 2012). Humanitarian crisis looms in Harare. Harare 24, Harare

Herald (December 15, 2014). Budget cuts detrimental to fighting climate change. Zimpapers, Harare

Herald (, 2013). Urban areas must plan for climate change. September 2 Zimpapers, Harare

Herald. (2016). Formulate Disaster Management Policy. April 01 Zimpapers, Harare

Herald. (2015). Beitbridge on the Mend. March 12 Zimpapers, Harare

Herald. (2016). Flood Hit Beitbridge. March 13 Zimpapers, Harare

Herald. (2016). Beitbridge Flood Victims Rise to 780. March 16 Zimpapers, Harare

Herald, (2016). Police Warn of Flash Floods. March 18 Zimpapers, Harare

Hodge, G. (2008). *Planning Canadian Communities: An Introduction to the Principles, Practice and Participants* (5th Ed.). Toronto, Ont.: Thomson/Nelson.

International Federation of Red Cross and Red Crescent Societies. (IDRL, 2010). *A rash of earthquakes shows that building codes can mean the difference between life and death.* E Newsletter Number 3, IDRL. Available online: http:www.bbc.co.uk/news/world-africa-20250494 [Accessed on 20 March 2016.]

International Federation of Red Cross and Red Crescent Societies. (IDRL, 2012). *Accra* tragedy shows that building code enforcement is critical to Disaster Risk Reduction. IDRL, Available online: http:www.bbc.co.uk/news/world-africa-20250494 [Accessed on 20 March 2016.]

IPCC. (2007). *Climate Change 2007: Impacts, Adaptation and Vulnerability. Working Group II Contribution to the Fourth Assessment Report of the Intergovernmental Panel on Climate Change.* Cambridge, UK: Cambridge University Press.

Kamusoko, C., Gamba, J., & Murakami, H. (2013). Monitoring urban spatial growth in Harare Metropolitan province, Zimbabwe. *Advances in Remote Sensing, 2*(04), 322-331.

Kreutzwiser, R. D. (1988). Municipal Land Use Regulation and the Great Lakes Shoreline Hazard in Ontario. *Journal of Great Lakes Research,* 14(2), 142–147.

Knutsson, P. (2006). The sustainable livelihoods approach: A framework for knowledge integration assessment. *Human Ecology Review, 13*(1), 90-99.

Leichenco, R. (2011). Current Option in Environmental Sustainability. *Elsevier.* 3(1)164 – 168

Listokin, D., and Hattis, D. (2005). Building Codes and Housing. *Cityscape: A Journal of Policy Development and Research,* 8(1), 28 – 67

Platt, R. H. (1998). *Planning and land use adjustments in historical perspective.* In R. J. Burby (Ed.). Cooperating with Nature: Confronting Natural Hazards with Land-Use Planning for Sustainable Communities. Washington, D.C.: Joseph Henry Press.

Rosenzweig, C., Solecki, W., Hammer, S., and Mehrotra, S. (2011). *Climate Change and Cities First Assessment Report of the Urban Climate Change Research Network*. New York, Cambridge University Press.

Richardson, R. (2012). *Land Use Planning Tools for Local Adaptation to Climate Change*. Available online:http://publications.gc.ca/collections/collection_2013/rn can-nrcan/M4-106-2012-eng. [Accessed on 10 June 2018]

Richardson, G. R. A., and Otero, J. (2012). *Land use planning tools for local adaptation to climate change*. Ottawa, Government of Canada.

Rosenzweig, C., Solecki, W. D., Hammer, S. A., and Mehrotra, S. (Eds.). (2011). *Climate change and cities: First assessment report of the urban climate change research network*. Cambridge University Press.

Shaw, R., Colley, M., and Connell, R. (2007). *Climate change adaptation by design: a guide for sustainable communities*. London, TCPA,

Spaliviero, M., Dapper, M. D. M., Mannaerts, C., and Yachan, A. (2011). Participatory approach for integrated basin planning with focus on disaster risk reduction: the case of the Limpopo river. *Water*, 3(3), 737-763.

Titus, J. G. (2011). *Rolling Easements. Washington DC: United States Environmental Protection Agency*. Available online: http:www.epa.gov/cre/downloads/rollingeasementsprimer.pdf. [Accessed on 10 June 2018]

TCPA. (2004). *Biodiversity by Design*, TCPA, London

Wilson, E., and Piper, J. (2010). *Spatial Planning and Climate Change*. Milton Park, Abingdon, Oxon; New York: Routledge.

Walsh, C, L., Hall, J, W., Street, R, B., Blanksby, J., Cassar, M., Ekins, P., Glendinning, S., Goodess, C. M., Handley, J., Noland, R., and Watson, S. J. (2007). Building Knowledge for a Changing Climate: Collaborative Research to Understand and Adapt to the Impacts of Climate Change on Infrastructure, the Built Environment and Utilities. Newcastle: Newcastle University.

World Bank, (2010). *Cities and Climate Change: An Urgent Agenda*. Washington DC: World Bank.

World Bank. (2012). *Building Urban Resilience: Principles, Tools and Practice*. Washington DC: World Bank

Zukunftsforum, Ö.S.t (2008). *Risiken und Herausforderungen ür die Öffentliche Sicherheit in Deutschland – Szenarien und Leitfragen*. Oktoberdruck GmbH, Berlin.

Zimstat, (2012). Census 2012 National Report. Government of
Zimbabwe, Harare.

Chapter 9

A Case for 'Retrofits' in the Urban Sector of Zimbabwe

Emma Maphosa, Innocent Chirisa & Zebediah Muneta

Introduction

The chapter seeks to explain how the retrofits projects can be adopted in the contexts of Zimbabwe particularly as an option in addressing climate change challenges such as water shortages, energy crises, increases in global warming and greenhouse gas emissions, flood management challenges and increase in waste management complications in all its sectors particularly the domestic sector residential sector. Retrofitting offers an opportunity to counter these challenges that have emanated as a result of rapid urbanisation and climate change among other factors by altering adding fitments to existing building stock. By dividing retrofits into water, energy, carbon and flood retrofits and through examples of specific such projects, the chapter examines the success of these retrofit initiatives in global wide, regional and local initiatives. Challenges examined in implementing these projects locally are examined. From the examination of these domestic projects and a comparison with successful retrofits projects elsewhere in the world, the chapters propose recommendations for action to make these initiatives a successful in the Zimbabwean context.

Retrofitting: A Review

Retrofitting is a concept that has recently gained prominence in the built environment world (Retrofit, 2050). Just like other concepts, this concept assumes a variety of definitions; Eames (2012) defines retrofitting as the direct alteration of the built environment fabric to improve energy, water and waste management efficiency of existing buildings or structures. It is therefore an opposite of reconstruction whereby the whole structure has to be demolished to come up with a new one that has a higher performance rate. However, these definitions all point to

163

the adjustment of parts of a built structure, partially, fitments to make it suitable to accommodate new emerging needs or a changing climatic context. This is equally in line with the building design principle of robustness. Retrofitting can be done also to make the building habitable especially residential neighbourhoods. It can also be done to improve the life span of structures and their values in the real estate world. From literature a variety of retrofits can be identified based on purposes that are energy retrofits, carbon retrofits, water retrofits, waste efficient retrofits and flood retrofits.

Energy Retrofit

Energy retrofitting refers to upgrading old structures so as to enhance their energy an environmental performance. This includes addressing issues of reducing water use requirements of a building, enhancing natural lighting, air and noise abatement characteristics of a building. Brown and Swan (2013) energy retrofits are fitments that are added to existing buildings to enhance their energy performance by reducing energy consumption as well as energy requirements for heating or cooling. Energy retrofits are technical devices that are fitted onto existing buildings to reduce energy use of appliances ad buildings fabric. Energy retrofits give emphasis to renewable energy retrofits such as photovoltaic cells onto buildings (Gelfand and Duncan, 2011). The aim of energy retrofits is to achieve he so called "near zero energy buildings" which are buildings with a high energy efficiency. Energy retrofits are particularly relevant in addressing the energy cycle imbalance of cities. Therefore, cities are failing to supply adequate energy as per their daily demand. Furthermore, assessments have to be carried out to determine the level of efficiency in terms of energy use of an area's building stock.

Water Retrofit

Water retrofits are structures fitted to conserve water or reduce the amount of water use required per building or household. This provides an opportunity given the present water challenges the world over. Given the water supply and management challenges in today's world alongside climate change reality, water shortages are becoming a common phenomenon in most neighbourhoods. While measures such as water policy regimes have been changed, participatory and integrated water resources programs are being taken, the contribution of the built environment itself also needs to take up an active role. Generally older buildings have built-in devices that consume more water per household. In order to reduce water consumption per household of older building stock, technically, water retrofits have to be introduced. Water retrofits can generally imply any devices fitted onto existing buildings to reduce their level of water consumption or water requirements to a level that is sustainable. These include devices that allow for wastewater reuse, low water flow, as well as a fitted rainwater collection system (Gelfand and Duncan, 2011). In addition, water retrofits can be fitted to solve water leakage challenges in the water supply appurtenances of old residential, commercial or industrial property to achieve water saving (Dixon, 2014) thereby the sustainable and green neighbourhoods as the United Nations Sustainable Development Goals.

Waste Efficient Retrofits are devices fitted on structures to enhance their waste management capabilities. For example, devices can be implanted in the water supply and sewage treatment works that separates solid waste by type that is whether they are recyclable or not. These enhance the recovery of waste for recycling and reduce the demand for space used as landfills within neighbourhoods. Though they are usually large-scale projects (Mixed waste MRF Retrofit, Sunnyvale Case Study, 2015). Waste retrofits are possible on individual buildings or blocks, they collect garbage, sort it into organic and degradable and non- biodegradable waste as well based on whether the waste material is biodegradable or not. Such technologies include the tri-sorter recycling system retrofit that separates waste into garbage container, garbage container and the recycling container (Waste Solutions Group, 2014). The organics and recyclable contents are then easily recovered and recycled which aids in waste management challenges.

Carbon Retrofit

Carbon retrofits are increment list improvements to an existing building fabric which are primarily motivated by the desire to lower the carbon emissions of the building fabric (Rhoads, 2010), tough in the process other benefits may accrue enhancement of environmental performance of building fabrics. The share of carbon emissions from buildings in the United Kingdom approximate to 45% (Carbon Action, 2050). This is driven by the desire to achieve zero carbon building stock. The low carbon development pathway in Zimbabwe that emphasises on development of renewable energy sources for domestic and transportation sector says little about addressing carbon issues at small scale levels. Carbon retrofits can be done through incorporating plants – greens on infrastructure to enhance carbon sequestration potential of buildings. Buildings contribute about 33% of greenhouse gas emissions.

Flood Retrofits

Flood retrofitting involve the structural adjustment of the building fabric ignored to reduce its susceptibility to flood damage risks and Therefore, providing safety from a social or human point perspective while from the business perspective this reduces property investment risks of property holders in flood prone zones. Unlike in the past when flood protection measures were designed as surface drainage systems and drains, today's planners are confronted with a land shortage crisis to the extent that as far as flood protection is concerned surface engineered approaches are less economic (Lammond *et al.* 2011). Therefore, other structure level approaches have been adopted to improve the flood resistance capability of structures in flood prone areas such as New York City. The approaches taken to minimise flood impact in this regard are referred to as flood retrofits. This includes wet and dry flood proofing technologies (Bloomberg, 2013). However, the studies note the relevance appropriate building codes. The most commonly mentioned retrofits are carbon and energy efficient retrofits in relation to their potential to curb the global climate change and energy crisis. These have attracted most discussion for their potential in line with the United Nations goal of creating sustainable cities by 2030. Therefore, retrofitting is a concept that tries to solve challenges by adapting the built environmental structures without reconstruction the challenges of which range from natural disaster (climate change, floods, droughts and heat waves), economic challenges of decline in property values, waste management challenge using cost effective and sustainable approaches. Retrofitting is therefore a concept that examines how the building and design world can solve urban challenges present and in future. Interesting to note is the consideration of the potential of retrofitting to address urban poverty issues (Williams and Gupta, 2013).

Retrofits projects have been conducted on large as well as small scale with variation in results for example the residential green retrofits in Germany, retrofits of suburbs in United Kingdom (Kelly, 2009), small scale energy retrofits in Cato Manor, Durban (Eskom Report, 2011) and the retrofit project in Greater Manchester (Hodson *et al.* 2012) where bricks of structures were changed from red to green. The benefits from these projects varied across scale and range from energy efficient savings in homes, employment stimulus, health and safety, reduced domestic waste production to a general improvement in quality of lives (Jonnes *et al.* 2013). Amidst these benefits literature also mentions challenges to retrofitting projects such as financial or manpower shortage, uncoordinated government policy, disintegrated approaches. This calls upon the requirements for a successful retrofit project which among other factors include; large-scale investment, engagement and stakeholder participation, enabling building policies and standards (Retrofit, 2050). From the study of literature little exists on retrofitting in African countries particularly Zimbabwe where it is faced with climate change challenge among other and is striving towards attaining the global agenda of sustainable cities. As far as retrofitting in relation to sustainability is concerned, the potential needs to be looked at. Furthermore, does large-scale retrofitting then have the potential to contribute towards addressing urban poverty and urbanisation challenges in Zimbabwe and Sub-Saharan Africa?

Urbanisation in Sub-Saharan Africa and the Rationale for Retrofits

As much as urbanisation is a global reality which is also a mixed bag that is it brings with it both positive and negative impacts, urbanisation in Africa and Sub-Saharan Africa is no exception. Therefore, urbanisation has already impacted negatively on the region's supplies of energy among other resources. In addition, it is also creating strains on the natural environment. Urbanisation has means population increase comes an increase in housing requirements, increase in demand for energy, increase in the demand for water for domestic, commercial and agricultural wastes as well as an increase in volume of waste generated within such communities. In terms of energy demand cities and the built environment are deemed to consume approximately 75% of the energy and at the same time emit 70% of the carbon emissions through the built environment (Eames *et al.* 2013). This shows that cities are unsustainable energy consumers amid the fact that

167

populations in these cities are increasing, the rate of energy consumption of buildings in cities is also likely to increase. In order to provide land for urban housing, the urban directly depends on land offered by the universe; to supplement energy requirements the environment is the source (water, coal and thermal fuels, solar). Likewise, for municipalities to provide water they have to be sufficient sources of streams, dams, underground and surface water sources. At the same time the same environment is used as a sink for carbon and other waste material generated by the increasing urban population. This therefore implies that as urbanisation rates increases, strain on the environmental services also increasing. With the current drive of building sustainable cities, we seek to examine how retrofitting as a strategy to make buildings accommodate the urbanisation challenges without compromising the welfare of urban natural environments.

Due to climate change impacts on renewable energy generation such as electricity, the future of electrical water supplies in Sub-Saharan Africa are threatened. Water levels are declining in its inland water bodies such as the Kariba Dam which is used for power generation for both Zimbabwe and Zambia which implies a loss in hydro generating capacity. This is amid the idea that 50-60% of the region's energy comes from hydro-power plants. Therefore, with an increase in climate change and droughts the region is likely to face severe supply shortages. Energy consumption in buildings depends not only on how they are designed or sited but how they are used by the occupiers as well (Jochem, 2010).

In addition to climate change population in Sub-Saharan Africa is increasing with over 50% of the population expected to be in cities by 2050. This implies that the need for more housing, commercial service provision in terms of education centres, health institutions among other services. Energy used in buildings account for 56% of the total energy consumption, in terms of energy consumption in buildings, Jochem (2010), states that residential buildings in Africa consume approximately 54% of the total energy consumption in other sectors. And the energy demand in Africa is estimated to be growing at 8% per annum amid reports of declining production capacity (Kitiyo, 2013). Then the option that remains is how the building sector adapts itself to the energy crises without risking the welfare of residents and its natural environment retrofitting offers the options.

Urbanisation in terms of cultural change and behaviours also impacts on the change in energy consumption in cities. As populations urbanise, they generally consume more energy relative to their counterparts. This can be explained by examining the energy consumption of the affluent populations in European Union countries and that of some Sub-Saharan Africa. According to Jochem (2010), populations in affluent regions consume 5- 10 times the amount of energy consumed in non-affluent regions of the world notwithstanding the fact that some of these regions experience extreme cold and therefore most of the energy is used for cooling or heating (Jochem, 2010). Likewise, as the majority of the population in Africa continue to urbanise demand strains on energy supply are inevitable.

An example of a successful retrofitting programme can be drawn from South Africa's Kuyasa state subsidised 30square meter housing units. It's a low-cost housing scheme. A retrofitting project was carried out to improve energy efficiency by promoting the use of renewable solar energy. Where solar water heaters (3KW input power and aluminium foil laminate were installed on 10 units. The project is 21 years and is expected to serve 2300 households from energy costs (Federal Ministry for Environment, Nature Conservation and Nuclear Safety, 2010). Other than the pressure of urbanisation on energy demand and use, rapid urbanisation in African cities is likely to threaten water availability in urban regions. Water is a precious commodity across the globe particularly in drier parts of the region. Therefore, design at every level should factor in water conservation approaches when designing buildings since the design of a building directly and indirectly influences water consumption for activities within the building. Water retrofits therefore are appurtenances that are fitted onto existing buildings ignored to enhance their water saving potential or allow the use of rainwater. An example can be drawn from the case of TIholego Eco village in South Africa.

Rooftop gutter systems collect rainwater from rooftops and channel it to underground water storage tanks. The water is then drawn and utilized for domestic and irrigation purposes during the dry season. Likewise, the designs of houses in this village have a fitted in grey water recycling system at household level. This allow for the collection of grey water from bathrooms, sinks which is collected in a separate storage tank and is available for secondary uses such as irrigation of lawns, vegetables. This option offers a

169

cheap method of ensuring water security within not only residential areas, but the example can be emulated for commercial and industrial buildings as well to improve their water use by retrofitting water tanks and wastewater collection systems onto the existing building frames (Federal Ministry for Environment, Nature Conservation and Nuclear Safety, 2010). Case of North Wood Residential Estate in West London retrofitting the suburb reduced the energy bills for residents (Heldgaard, 2012). Not only were benefits limited to reduction of energy expenditure but comfort from temperature increase nuisances and value of the property within the estate also rise and attracted more tenants into the estate.

Zimbabwe and Retrofits

With increasing population growth in urban areas of Zimbabwe and the increasing demand for services such as water, energy and electricity of which a greater part of it is generated from hydroelectricity and thermal power stations Harare and the Harare thermal power station, Masvingo still depending on only hydro-electricity crises in the urban areas in the near future are inevitable. The urban sector generally consumes more resources than it replenishes indicating that as counties continue to urbanise demand for ecosystem services will also rise. The building sector (housing, industry, agriculture and commercial sectors are the among the major energy consumers. Given the energy consumption for residential property increases as they age, it is likely that the current building stock in cities will cost much of the nation's energy bill. In addition, amid climate change realities and water crises the current design of the building stock is likely to heighten the level of water and energy consumption among other factors. Therefore, the calls to readjust the current building elements to reduce energy consumption and contribute towards mitigating impacts of the built environment on climate change through retrofitting. While all sectors of the urban sector should retrofit, emphasis should be on the housing sector since it is by far the largest energy consumer in Zimbabwe according to the SADC/RERA (2013). According to the report electrical energy consumption in the housing sector rose by approximately 1.4% per annum between 1996 and 2006, this was above all the other sectors industry, agriculture. In contrast, energy consumption in industries has generally declined during the same period.

Urbanisation is the general rise in % of a nation's total population in cities relative to that in rural areas. Since the dawn of independence in Zimbabwe that is the period 1980 and thereafter, the rate of urbanisation in Zimbabwean cities has been gradually increasing. Since 2010 the rate of urbanisation is approximately increasing at 3.4% per annum (Mugumbate and Chamunogwa, 2013) (Zimstat, 1992, 2002, 2012). This has been attributed to the city's open-door policy following the removal of the restrictive colonial pass system. Therefore, populations have flocked into cities with Harare, Bulawayo containing the largest growth in population. It is no doubt the pace of rural- urban migration brought with it parcel of requirements to the urban ecological system of the cities among which include the demand for more energy (both for domestic, transport and industry), demand for water for the same rationales, increasing demand for infrastructure (housing, industry, public service institutions, recreational space that would drain into the natural landscape). Apart from the increase in demand for ecological services, the populations and human economic activities in turn produced by products such as solid waste products, municipal wastes, carbon dioxide which require deposition and, in most cases, open environments have been taken as the cheap sources for disposal. Therefore, with an increase in population, demands for urban ecosystem services are likely to increase and in turn the impacts of human activities on the natural ecosystem. This is against the background that the urban areas are already under stress to provide ecosystem services (water, energy, clean air, recreational space) and the fact that the local climate is already changing. Therefore, the sustainability of urban ecosystem therefore is questionable. Hence retrofitting appears to be among the options that remain to be exploited in Zimbabwean cities to address these foregoing challenges under the lead of the building sector.

Currently, energy markets consist of housing, industry, agriculture, commerce and transport. Energy management is under the Ministry of Energy Power and Development. The 2012 report by the ministry indicated that energy production from its combined sources hydro power, coal power, biogas stations and biomass (mainly wood fuel in rural areas) is failing to meet current demand with a shortage of 34% (Ministry of Power and Development Report, 2012). In the case of biomass supplies the report indicates that the rate of use is unsustainable given consumption rate of 6 million tonnes per annum against a tree planting rate of around 10 million tonnes per annum. This shows that six years are required to replenish the quantity exploited during a single year which is unsustainable. Hydro- electric supplies are also under threat of declining water levels in water bodies such as the Kariba Dam which is its major supplier as a result of climate change. While there are positive reports of coal reserves, the challenge stands that combustion of coal releases carbons which depletes the ozone. Therefore, it accelerates the rate of climate change. This shows that the country's supplies of energy are threatened.

Deterioration of buildings in the CBD and low occupancy rates have been widely reported in media reports (Kavaza,11January 2014). Deterioration of structures has an impact on real estate market returns and also threatens human health and the physical environment. Though other factors can be pointed to the vacancy of Harare CBD Buildings such as declining economic activities and a generally declining economy, retrofitting them could offer a cheaper alternative. Water Security threats from climate change, contamination of water bodies Nhapi *et al*. 2011) and a general decline in watershed areas for cities due to competition in demand for space either for housing, industry or commercial use. This then threatens the viability of water supply in Zimbabwean cities water consumption in houses, towns and industry. Water utility fitments in structures include water sinks, bathtubs and toilets.

Increasing populations in Zimbabwe's cities imply an increase in solid waste generation amid reports of a declining economy and increasing land shortages therefore point to a scenario where buildings have to have built-in systems of waste recycling and management. This can be seen by widespread dumping sites in cities include the Pomona Dump Site in Harare which is already used up (*NewsDay*, 17 February 2012). Though new sites are sought the sustainability of this waste disposal approach particularly amid competition for land, is questionable. In this case waste efficient retrofits in building structures could reduce the volume of solid waste generated by households or industry and in so doing help in preserving the natural ecology of the city. However, with an increase in population and the demand for land, the sustainability of using dump sites remains to be questioned. However, options have to be found which are cognizant of which retrofitting structures to reduce waste generation is one.

As much as Zimbabwe's development paradigms are filled with "affordability" agendas-affordable housing (Moyo, 2014), affordable water services, affordable energy, retrofitting can form part of the affordable development package. This is because the degree of energy consumption in a home has cost implications to a household particular during the period of its use. In general, the higher the energy use the higher the rental charges and therefore, the higher the cost of running the home. Though it should be noted that energy use depends on socio- economic characteristics of the household, the nature of the building design and its fixtures are to large extent liable for energy wastage. Therefore, the ability to reduce energy requirements of buildings during their operation life has a great potential to save tenants from unnecessary energy costs in so doing increasing the affordability of property ceteris paribus.

Retrofitting can be necessary to offset the impacts of population growth strains on ability of building particularly residential compartments. Population increase can result in deterioration of quality of services offered by a building. In addition, a shift in population composition may imply that housing units have to be retrofitted to cope with the new demands otherwise their defects in the structure cost the environment or welfare of the building users. A typical case is Mbare Housing flats in Harare. It is a government housing scheme south of Harare. Originally the housing flats were designed as single men, migrant worker quarters, however with the open-door policy to the city following independence, there has been an influx of population into the city and consequently therefore demand for housing of which Mbare flats were converted into family use homes. Therefore, the ability of such flats to accommodate of the family is questionable (Dialogue on Shelter, 2013). This may also point to the need for retrofitting adding elements and facility that make the blocks family friendly to which plans are underway, (Mashava, 3 October 2015) Emphasis in this discussion however is directed towards retrofitting the flats to ensure energy efficiency, water efficiency amid reports of critical shortages of such infrastructure or where the infrastructure exists it uses the traditional approaches.

Discussion

In the discussion it can be seen that in various challenges are incurred in retrofitting older buildings to respond to the new challenges such as energy crises, water shortages and increasing air pollution through release of carbon. These challenges range from financial shortages (African Development Bank, 2010; Figure 9.1), uncertain political environment and supply driven projects with little input from demand or user capacity and views and therefore fail upon implementation. A typical example is the use of the prepaid water meters in household water units to reduce the risk of non- revenue water in cities such as Harare, Mutare and Rusape among others. The initiative has been widely objected as a move to benefit only local municipalities and reduce access of water as a basic right for everyone particularly the poor (Gambe, 2015; Nhema and Zinyama, 2016). In terms of solar water geysers their implementation is at a minor scale and on optional bases. Could they have been made mandatory, their impact in terms of increasing energy efficiency and reducing carbon production could have been great? This represents a shortfall on backing legislation. There is scarce information on flood retrofit projects in urban cities, yet a majority of its uprising so called "informal settlements" are mostly located in low lying and flood prone areas that have been preserved from settlement (IIED, 2008).

Table 9.1: Retrofits Specific to Urban Zimbabwe

Type of Retrofit	Specific Developments in Zimbabwe Urban Areas	Comments
Energy Retrofit	National solar water heating project (currently, replacing electric geysers with solar powered geysers) (Harare, Bulawayo, Gweru, Masvingo (Hove *et al.* 2007)	Lack of capital investments (Hove *et al.* 2007), costly for domestic markets and institutions- hospitals and boarding schools., 89% of institutions requiring an increase in hot water supplies.
Water Retrofit	1.WASH small towns programme (Bindura, Zvishavane) Urgent water supply rehabilitation and sanitation project (6 towns-Harare, Mutare, Chegutu, multi donor trust fund— (African Development Bank, 2010) 2.prepaid water meters (to detect water leakages and	Urgent water project- weak institutional coordination, resource shortages, political situation of the country (African Development Bank, 2013) - Residents objecting against the initiative - Human right activist against the initiative arguing it is not- pro-poor. - The solution is demand driven, taken by suppliers (municipalities), without sufficient consultation of the market

	reduce level of non-revenue water losses - Mutare, Rusape, pilot project Harare	- One sided approach advantageous to municipalities in terms of dealing with challenges in payment of water rates and illegal abstractions (Gambe, 2015). - Misconceived initiative in terms of its ability to increase efficiency in use of water in buildings.
Carbon Retrofit	Solar panels, renewable energy -insulation of roofs	-implemented on a minor scale on voluntary basis - Should be supported by regulation-made mandatory.
Flood Retrofit	-rare literature on such projects (structural elevations and dry flood proofing) (they could have been occurring at individual level)	-rare information pertaining to such projects. Only managerial and non-structural approaches have been taken to improve emergency awareness in times of floods by the Civil Protection Unit in January 2008 flood in Harare and Chitungwiza. -No statutory regulations or design standards for buildings in flood prone areas (flood codes- Muyambo and Klaasen, 2015). -most of the structures in low –lying areas are informal settlements therefore reluctance to establish flood resilient approaches for such settlements- areas along the Mukuvisi

Conclusion, Policy Options and Way Forward

Since there is need to secure a secure financial base before such projects can be implemented since financial shortages have been cited as a challenge in Wash Water Rehabilitation Project. Options such as partnerships with the Green Infrastructure Finance can be exploited to improve finance for large-scale retrofit projects. In addition, retrofit projects should be demand-driven, if they are driven by demand, small scale and from the local initiatives they can be more sustainable since the locals can manage the costs of maintaining the functionality of these initiatives (as lack of capacity to install and manage these retrofits has been accredited as a challenge. User engaged projects are also readily acceptable by users upon inception. In addition, building legislation and standards should take into account the view that solar energy fitting should be mandatory for every development. Unless backed by law, the initiatives remain rare and their benefits cannot be fully appreciated. Whilst these factors are important in implementation of retrofit projects, actors in the development market have to be engaged to appreciate the relevance of a move towards energy, water and flood conscious designs.

References

African Development Bank. (2013). Urgent Water Supply and Sanitation Rehabilitation Project; Phase 2, African Development Bank Available online: https://www.afdb.org/fileadmin/uploads/afdb/Documents/Project-and-Operations/Zimbabwe_Urgent_Water_Supply_and_Sanitation_Rehabilitation_Project_-_Phase_2_-_Appraisal_Report.pdf. [Accessed on 17 June 2018]

African Development Bank. (2010). Urgent Water Supply and Sanitation Rehabilitation Project: Project Appraisal Report, African Development Bank Available online: https://www.afdb.org/fileadmin/uploads/afdb/Documents/BoardsDocuments/ZIMBABWE_AR_-_UWSSRP.PDF[Accessed on 17 June 2018]

Bath., and North- East Somerset Local Development Framework. (2013). Sustainable Construction and Retrofitting; Supplementary Planning Document. Available online: http://www.bathnes.gov.uk/sites/default/files/sitedocuments/Planning-andBuildingControl/PlanningPolicy/LP20162036/lp_sa_scoping_report_annex_a.pdf. [Accessed on 17 June 2018]

Bloomberg, M. R. (2013). *Coastal Climate Resilience; Designing for Flood Risk*, New York

Dialogue on Shelter. (2013). Harare Slum Upgrading Profile, 1-16. Available online: http://www.dialogueonshelter.co.zw/component/content/category/16-projects.html. [Accessed on 17 June 2018]

Dixon, T. (2014). 'City-wide or City-blind' An Analysis of Emergent Retrofit Practices in the UK Commercial Property Sector, Retrofit 2050, Working Paper 1, 1-87

Dixon, T., and Eames, M. (2013). Scaling Up; Challenges of Urban Retrofit, *Building Research and Information*, 41 (5), 499-503

Eames, M., Dixon, T., May, T., and Hunt, M. (2013). City Futures; Exploring Urban Retrofit and Sustainable Transition, *Building Research and information*, 41 (5), 504-516

Eskom. (2011). Eskom Factsheet 2013, Green retrofitting for Low-income Homes in Durban. Available online: http://www.eskom.co.za/OurCompany/SustainableDevelopment/ClimateChangeCOP17/Documents/Green_retrofit_of_low-income_homes_in_Durban.pdf. [Accessed on 17 June 2018]

Federal Ministry for Environment, Nature Conservation and Nuclear Safety. (2010). Sustainable Buildings and Construction in Africa, Federal Ministry for the Environment, Nature Conservation and Nuclear Safety, Johannesburg

Gambe, T. R. (2015). Prospects of Prepaid Smart Water Metering in Harare, Zimbabwe, *African Journal of Science Technology and Development*, 7, 236-246

Gelfand, L., and Duncan, C. (2011). *Sustainable Renovations, Strategies for Commercial Building Systems and Envelope*, Willey, Chichester

Heldgaard, T. (2012). The Social Implications of Energy Efficient Retrofits in Large Multi-Storey Tower Blocks, LSE Housing and Communities, Case Report 75, 1-60

Habitat, U. N. (2012). Sustainable Urban Energy, A Sourcebook for Asia. *Nairobi, Kenya: United Nations Human Settlements Programme (UN Habitat).*

Hove, T., Mabvakure, B., and Schwazlmuller, A. (2007). Final Report on the Survey on Demand of Solar Water Heaters in the Institutional Sector, Zimsun Market Survey 200-2007, 1-54

Hodson, M., Thompson, M., and Marvin, S. (2012). Retrofit and Greater Manchester; Landscape, Governance and Practice, Working Paper Retrofit 2050

International Institute of Environment and Development Briefing (IIED). (2008). Against the Tides; Climate Change and High-Risk Cities

Johanes, P., Lannon, S., and Patterson, J. (2013), Retrofitting Existing Housing; how far, how much? *Building Research and Information*,. 41 (5), 532-550

Kawadza, S. (11 January, 2014). Harare's Dens of Lions, *The Herald*, Available online: https://www.herald.co.zw/harares-dens-of-lions/. [Accessed on 17 June 2018]

Kelly, J. M. (2009). Retrofitting the existing UK Building Stock, *Building Research and Information*, 37 (2), 196-200

Lammond, J., Booth, C., Hammond, F., and Proverbs, D. (Ed.) (2011). Flood Hazards; Impacts and Responses for Built Environment, McGraw Hill, New York

Lockwood, C. (2009). Building Retrofits, Urban Land (1-12) Available online: https://www.mdpi.com/2075-5309/4/4/683/htm. [Accessed on 17 June 2018]

Mercader-Moyano, P. (Ed.). (2015). *The Sustainable Renovation of Buildings and Neighbourhoods*. Bentham Science Publishers.

Mashungu, L., and Kavhu, S. (2016). Zimbabwe's Nationally Determined Contributions Renewable Energy Developments; Regional Expert Meeting on Climate Change and Enhanced Renewable Energy Deployment in East and Southern Africa, Addis Ababa

Mushava, S. (3 October, 2015). Whither Mbare Flats? *The Herald*, Available online: https://www.herald.co.zw/whither-mbare-flats. [Accessed on 17 June 2018]

Moyo, W. (2014). Urban Housing and Its Implication on Low-income Earners of a Harare Municipality, Zimbabwe, *International Journal of Asian Social Sciences*, 4 (3), 356-365

Ministry of Power and Energy Development; National Energy Policy Zimbabwe

Muyambo, A., and Klaasen, W. (2015). Capacity Building for Zimbabwe's Urban Local Authorities in Water Supply Needs Assessment, Netherlands Water Partnership

National Disaster Management Authority Government of India. (2014). Natural Disaster Management Guidelines; Seismic Retrofitting of Deficient Buildings and Structures, New Dehli

Nhema, A. G., and Zinyama, T. (2016). Prepaid water meters in Zimbabwe: the quest for an efficient water delivery and cost recovery system. *Global Business and Economics Research Journal*, 5(1), 1-30.

RERA (Regional Electricity Regulators Association of Southern Africa). (2013). Supportive Framework Conditions for Mini-Grids; Employing Renewable and Hybrid Generation in the SADC Region; Zimbabwe Case Study; Gap Analysis and National Action Plan, RECP,

Rhoads, J. (2010). Low Carbon Retrofit Toolkit—A Roadmap to Success. Available online: http://climatechangeecon. org/index. php. [Accessed on 17 June 2018]

Swan, W., and Brown, P. (2013). *Retrofitting the Built Environment,* Wiley, Chichester

UNDP (2012), Brown Bag Dialogue Series 2, 1-7, Available online: https://info.undp.org/docs/pdc/Documents/ZWE/2013%20 Q1%20Workplan%2000079955.pdf. [Accessed on 17 June 2018]

United States Green Building Council Available online: https://www.usgbc.org/articles/green-building-facts [Accessed on 17 June 2018]

Waste Solutions Group. (2014), Tri-sorter Recycling System Retrofit, Available online: https://www.acmo.org/wpcontent/uploads/CMMagazineArch ives/2008/CMwinter2008%E2%80%8B/winter2008e.pdf. [Accessed on 17 June 2018]

William, K., and Gupta, R. (2013), Retrofitting England's Suburbs to Adapt to Climate Change, *Building Research and Information,* 41(5), 517-531

World Bank. (2015). Stock taking of the Housing Sector in Sub-Saharan Africa; Challenges and Opportunities, World Bank, Washington DC

Chapter 10

Globalising International Environmental Law in Zimbabwe: Practices, Gaps and Direction

Halleluah Chirisa, Aurthur Chivambe & Liaison Mukarwi

Introduction

The concept of Integrated Environmental Management (IEM) is neither well-understood nor systematically practiced. The concept of IEM in Zimbabwe is managed through the Environmental Management Act (EMA) of 2002. One of the major objectives of the Act is to promote the sustainable management of Zimbabwe's natural and physical resources. One of the major environmental conflicts in Zimbabwean society is the debate between the white population and traditional Africans over the preservation and management of the extensive forests that cover approximately 18% of the country (Moyo *et al.* 2014; Sunderlin *et al.* 2005). Whites overwhelmingly want such areas to be preserved in their natural state. The majority of the African population living at the subsistence level, however, wants to cut the forests for two reasons. First, for most Africans, wood is the only readily available energy source. Second, many want to transform the forest into farmland in an attempt to mitigate the steady population growth in the black community (Leach and Mearns, 2013). Zimbabwe's Ministry of Environment and Tourism is responsible for the administration of eleven national acts (Sandler *et al.* 2012). These current statutes and instruments used include the EIA and Ecosystems protection Statutory Instrument (SI) 7 of 2007, Solid Waste Management SI 6/2007; SI 98/2010, Atmospheric pollution control SI 72/2009, Hazardous substances and waste management SI 12/2007; SI 10/2007; SI 77/2007 AND Effluent Dispersal Control SI 6/2007. Some of these acts relate only indirectly to environmental protection, however and there exist several other environmental acts that do not fall under the purview of the Environment Ministry (Moyo *et al.* 2014). In fact, a total of six other ministries are engaged in environmental protection. This overall administrative structure may be the result of several different factors. Zimbabwean authorities may believe that environmental issues are best addressed on a broad regulatory spectrum. In reality, as a practical concern, available funding (both international and domestic) may be specifically targeted for certain areas of concern. It would make sense; therefore, to keep environmental legislation broadly dispersed (Wunder *et al.* 2008). The concept of environmental impact assessment has gained only informal recognition in Zimbabwe (Nicholas, 1994). For instance, most agencies and commissions will consider a variety of environmental impacts when making decisions, but no legal mandate exists requiring that a specific process be followed. Moreover, citizen enforcement provisions are rare in Zimbabwean legislation. Several commentators have suggested that if Zimbabwe wishes to address the concerns of indigenous populations in national policies, citizen suit provisions could facilitate this aim (Brunch, Coker and Van Arsdale, 2001).

This present study is an attempt to evaluate what Zimbabwe has been doing in practice, identifying the gaps and proffer a nuanced and informed direction toward the protection of not only the global commons but also the local natural resources in a sustainable manner. This study has used desktop document review, forecasting and back casting as methodology methods. This is so because forecasting helps to shape the environmental law implications in the progressive future and through back casting one can be able to put environmental policies based on history, conferences and workshops of the past. For example, The Paris Agreement on climate change, the Rio Earth Summit of 1992 and the Johannesburg Summit of 2002 to mention a few.

Background of the Study

The issue of environmental protection is a global phenomenon which has transcended to be a regional and local issue. Global conventions and treaties set the tone for environmental protection with regional blocks and member states translating the pronouncements into action in line with their contexts and resources at hand. This section presents a background of the issues of international environmental law viewing it from two angles, the global context and regional context.

Global Context

Most states, globally, have braced themselves for the tough task to protect the environment, with the developed world taking a lead in these efforts. It has been acknowledged by the United Nations that the developed world has better capacity to promote environmental protection, in form of human capital, financial resources and technologies, than the developing world. In this sense, it urged the developed world to transfer and share these capacities with the developing countries so as to promote a sustainable development globally. Though actions towards environmental sustainability in respective developed nations vary, there is consensus that they are better positioned than their counterparts; they have effective legislations and enforcement mechanism that truly uphold the internationally recognised standards as spelt in the international environmental law. For example, in 2015, China began to implement its updated 1989 Environmental Protection Law (EPL), which suggests that China has become more serious about improving environmental quality (Zhang, 2012). The most significant additions and provisions to the EPL include: (1) increasing the seriousness of the consequences for violating China's environmental laws (2) expanding the scope of projects subjected to environmental impact assessments and (3) allowing nongovernmental organisations to take legal action against polluters in the public interest (Zhang, 2012 and Shinn, 2016). In 2016, The China is being one of the world's biggest emitters of greenhouse gases, formally joined other states including USA to formally ratify the Paris climate change agreement in the battle against global warming. The Paris Agreement commit the member states to aspire to keep temperatures below 1.5C above pre-industrial levels, by reducing GHG emissions and for rich countries to continue giving climate aid to poorer countries beyond 2020.

The Environmental Performance Index (EPI) indicates that eight of the top 10 countries are found in Europe, with Singapore and Australia completing the top of the ranking. EPI is an attempt to measure the extent to which countries implement policies to protect their natural environments, by calculating 20 separate indicators across areas such as air quality, agriculture, climate and energy (Morgan, 2012). For instance, Singapore's green efforts can be traced back to the late 1960s when the country was undergoing rapid industrialisation and urbanisation. It has Green Plan of 2012, which is Singapore's ten-year plan for achieving sustainable

183

development (Mbula, 2016). It describes the strategies and programmes that Singapore would adopt to maintain a quality living environment while pursuing economic prosperity. It also contains a list of specific targets that need to be met (*ibid.*). Also, as an example, the Australian Government has a range of environmental policies to minimise the impact of government operations on the environment. There are also agency measures and targets for carbon emissions, energy, waste and resource use, as well as set mandatory environmental standards for incorporating sustainability into government procurements (Slaughter, 2009). Even though the developed world takes the lead in protecting the environment today, the UN recognises the historical responsibility of the developed countries in causing global warming even though current industrial activity in major developing countries such as China and, to a much lesser extent, India is adding incrementally to that stock (Zhang and Crooks, 2012). This gives an impression that if developed countries do not make significant and absolute reductions in their emissions there will be a progressively smaller carbon space available to accommodate the development needs of developing countries (*ibid.*).

Regional Context

Many developing countries and countries with economies in transition are attempting to strengthen and consolidate their environmental management systems (Lee and George, 2000). These reforms are part of broader trends in global environmental consciousness that are taking place in the era of political and economic change through the processes of globalization. Globalization often accelerates economic growth and increases environmental deterioration in developing countries and therefore, has important implications for the use and development of environmental management statutes (Morgan, 2012). These effects and implications, in turn, differ across and within developing regions and countries depending on their level of development, dependence on natural resources and other factors. Tanzania appears to be the first African nation in which courts have addressed the scope of constitutional right-to-life provisions in the context of environmental protection (Brunch, 2001). Article 14 of Tanzania's constitution provides that everyone has the right to exist and to receive from the society protection for his life, in accordance with the law. For instance, In *Kessy*, citizens of Tabata, a suburb of

Dar es Salaam, brought suit against the City Council of Dar es Salaam, seeking to enjoin the city from operating a garbage dump that created severe air pollution in nearby neighbourhoods. The foul smells and air pollution caused respiratory problems to the residents; children, pregnant women and the elderly suffered the most. The citizens won a judgment in 1988 in which the court ordered the City Council to cease using the Tabata area for dumping garbage and to construct a dumping ground where it would pose no threat to the health of nearby residents (Environmental Law Institute, 2007).

In 1997, Ethiopia approved its first comprehensive environmental initiative and subsequently implemented strategies and laws to support sustainable development (Bekele, 2008). Ethiopia enacted a wide range of legal, policy and institutional frameworks regarding the environment; water, forests, climate change and biodiversity (Damtie and Bayou, 2008). Ethiopia signed a number of international environmental treaties and ratified the global environmental conventions. Also, Mali has relatively well-developed environmental legislation, having initiated a political and institutional framework dealing with climate change. In 2010, it established the National Agency for Environmental and Sustainable Development, which was responsible for implementing environmental policy and integrating bureaucratic responses. Mali also has a strategy for a green economy and climate change. In 2011, the World Bank improved Mali's score for environmental policy and institutions. Taking Zimbabwe, it has its Constitution awarding environmental rights to its citizens. To action its commitment to international declarations and treaties it ratified, the country has numerous environmental laws that are in a way capturing the emerging trends in Environmental protection for example the EMA Act makes it mandatory for an EIA on all listed developments before commencing.

However, the environmental initiatives in Africa and most developing countries have suffered a still birth due to slow policy implementation, poor legal enforcement and insufficient financial and human capacity to carry out the agreements. While their environmental laws are sometimes impressive, implementation is often lacking. It has also been difficult to integrate environmental initiatives into national development plans and poverty reduction strategies (The Environmental Management Act (EMA), 2011). African governments have signed many treaties and agreements but

185

have generally failed to articulate coherent solutions to their environmental problems. As an example, a coup d'état in 2012 in Mali resulted in a ninety percent reduction in the budget of the Ministry of Environment and Sanitation's environmental department and this affected efforts in environmental managements. In addition, Mali has low regard for environmental legislation and weak human capacity for environmental improvement. Despite the fact that Ethiopia takes environmental protection more seriously than most African countries do, it continues to suffer from environmentally related problems due to inadequate implementation and enforcement of environmental laws. Pollution monitoring, reporting and verification of abatement measures have been weak (Damtie and Bayou, 2008). Lack of human capacity has been the main constraining factor. Also, Zambia has no constitutional recognition of environmental issues; its constitution does not have an explicit requirement for environmental protection (The Constitution of the Republic of Zambia, 1996). Absence of an explicit constitutional requirement for environmental protection or at least a guarantee of the right to a healthy and clean environment strongly suggests that Zambia is still lagging behind in terms of executing its obligations under international law. In general, the approach of individual African countries towards protection of the environment varies enormously. Some countries have impressive legislation in place, while others lag behind. Even in the case of countries with a relatively strong commitment to the environment and reasonably good legislation, there are serious shortfalls in funding and human capacity to implement programs to protect the environment. Overall, most African states have weak bureaucracies.

Theoretical Perspectives

Protecting the environment is a theme that runs through many of the new Global Goals for Sustainable Development, from combating climate change to safeguarding the world's forests and oceans. The debate on integrated environmental planning stems from the externalities of development which include degradation/pollution of the environment (Mohamed-Katerere, 2001) Sustainable development is quite vital to integrate several environmental and economic goals that meet the needs of tomorrow as well as those of today in planning. The World

Conservation Strategy (WCS), the World Commission Environment and Development, the Brundtland Commission (WCED, 1987) and prompted closer links between the environment and development (Arts, 2009). The World Conservation Strategy emphasized the integration of the environment and conservation values/concerns into the development process and the WCED emphasized issues of several and economic sustainability. In a study that was conducted by Axford (1997) he found out that the major barriers to integrated environmental management were; inadequate information, lack of advocacy of an environment vision and leadership, resistance to change, political interference, lack of resources and the lack of clarity between regional and territorial boundaries.

International Environmental Laws

In general, environmental laws are the standards that governments establish to manage natural resources and environmental quality (Biswas and Agarwal, 2013). The broad categories of natural resources and environmental quality include such areas as air and water pollution, forests and wildlife, hazardous waste, agricultural practices, wetlands and land-use planning (Rajah *et al.* 2012). International environmental law is a body of international law created by states for states in an attempt to control pollution and the depletion of natural resources within a framework of sustainable development (*ibid.*). Environmental law, at both the national and international level, has two basic rule types: those designed to ensure compliance or conservation, these are prescriptive and those designed to facilitate better practice, these are process oriented (Mohamed-Katerere, 2001). Various aspects of governance are now addressed in law, using both rule types, in multi-lateral environmental agreements, including public participation, access to information and due process. Not all of these create rights that are actionable against the state. Nevertheless, they impose a duty on the state to develop national legal systems that recognise such rights (Mohamed-Katerere, 2001). In some instances, it encourages the development of private law rights such as traditional resource rights. These developments have occurred within the context of a growing body of human rights law dealing with development, justice, fairness and equity.

The environmental law can be in form of declarations or treaties ratified by member state in the world. There are two major declarations on international environmental law which are; the

Declaration of the United Nations Conference on the Human Environment (the 1972 Stockholm Declaration) (UN, 1972) The 1972 Stockholm Declaration represents the first major attempt to consider the global human impact on the environment and an international attempt to address the challenge of preserving and enhancing the human environment. The Stockholm Declaration espouses mostly broad environmental policy goals and objectives rather than detailed normative positions (UN Habitat, 2014). The Rio Declaration on Environment and Development produced at the 1992 United Nations Conference on Environment and Development (UNCED), known as the Rio Earth Summit consists of 27 principles intended to guide future sustainable development around the world. The list of environmental law treaties among others, include:

- Vienna Convention for the Protection of the Ozone Layer, 1985,
- Montreal Protocol on Substances that Deplete the Ozone Layer, 1987
- Basel Convention on the Control of Transboundary Movements of Hazardous Wastes and their Disposal, 1989
- Convention on Biological Diversity, 1992,
- Cartagena Protocol on Biosafety to the Convention on Biological Diversity, 2000
- United Nations Framework Convention on Climate Change, 1992 (UNFCCC)
- Kyoto Protocol to the United Nations Framework Convention on Climate Change, 1997
- United Nations Convention to Combat Desertification in those Countries Experiencing Serious Drought and/or Desertification, Particularly in Africa, 1994 (UNCCD)
- Convention on the Law of the Non-Navigational Uses of International Watercourses, 1997

In the realm of international environmental law, there is soft and hard international environmental law (Brunch *et al.* 2001). A distinction is often made between hard and soft international law. Hard international law generally refers to agreements or principles that are directly enforceable by a national or international body. Soft international law refers to agreements or principles that are meant to influence individual nations to respect certain norms or incorporate them into national law (*ibid.*). Soft international law by itself is not enforceable. It serves to articulate standards widely shared, or aspired to, by nations. Despite having hard international environmental law, declarations and treaties, there is no international court for the environment (Arts and Buizer, 2009). Environmental disputes have been litigated before a wide range of adjudicative bodies - global and regional, judicial and arbitral (*ibid.*). Many multilateral environmental regimes have 'non-compliance procedures' which are typically non-judicial.

Contextualising International Environmental Law

International cooperation in the form of treaties, agreements and resolutions created by member states as well as national laws and regulations are being used to protect the environment (Environmental Law Institute, 2007). Since ultimate responsibility for the protection of the environment remains at the national and local level, the laws and regulations related to the environment should be locally borne to embrace the local context. Context refers to the combination of circumstance that determines the norms, standards or thresholds for environmental sustainability performance (Brunch *et al.* 2001; Bekele, 2008). Debates over equity in the context of global environmental problems have been reflected in the development of an array of measures designed to take into account the special positions of all countries especially the developing countries (Brunch *et al.* 2001). These can take form the form of differential treatment where different groups of countries take different commitments or where the implementation of developing countries' commitments is conditioned upon the implementation by developed countries of their financial and technology transfer commitments (*ibid.*). This has also taken form of differential measure such as the introduction of contextualising commitments that apply to all countries in the same way but allow countries a level of flexibility in their implementation, depending on their specific situation (Biswas and Agarwal, 2013). Differential treatment in international environmental law is a pragmatic response that states have adopted in the search for ways to effectively address environmental problems (*ibid.*). It also contributes to the realisation of substantive equality and is therefore, an equity measure.

International environmental laws developed in large part in response to growing environmental degradation. Yet because environmental conservation is intrinsically connected with the use of environmental and natural resources, livelihoods and realisation of human rights, the necessity to adopt a broad local view of environmental regulation is very important (El-Fadl and El-Fadel, 2004; Cornwall and Nyamu-Musembi, 2004). This avoids the use one size fit all approach in environmental matters, but individual states must contextualise the global laws in light of the resources at their disposal for the task. A uniform approach to environmental regulations will disadvantage many of the developing states (Cornwall and Nyamu-Musembi, 2004). A just environmental system is based on formal equity versus equality. Equality can produce an optimal aggregate outcome, such as a high rate of overall economic growth but tends to overlook the welfare of disadvantaged individuals. In fact, equality of right or opportunities does not necessarily bring about equality of outcomes (*ibid.*). This is relevant in a world characterised by disparities in resources and capabilities, between individuals and between states. It is not sufficient to assert that like cases must be treated alike.

The realisation of substantive equality implies that existing inequalities such as inequalities in wealth or natural endowments should be acknowledged and taken into account when developing regulations if compliance is to be achieved (Rajah *et al.* 2012; Cho and Tifuh, 2012). Most African states have started politicising the environmental issues blaming the environmental obligations arising from global conventions and treaties of stifling development in their jurisdictions. Most African states are of a view that the developed world is using the environment to make them remain poor given that the environmental provisions pronounced by most the international laws has an economic impact, yet Africa has no economic muscles to bankroll the demands (Environmental Law Institute, 2007). As long as the people of Africa treat some of the laws as meant to stifle developments in their jurisdictions, they won't be fully committed to uphold these international laws. Despite the United Nations urging the developed world to transfer resources, human, technological and financial, to the developing world for environmental protection especially reduction of GHGs, the efforts are still low, and some states view that as political patronage where they have to be a puppet of such nation providing aid (Kalima, 2002). African countries that value sovereignty very much are victims, they benefit little from those resources and therefore, their efforts to uphold international environmental laws remains compounded (Damtie and Bayou, 2008).

Results and Discussion

Zimbabwe is a vibrant player at various levels in the international and regional environmental arena (Mkandla, 2014). It acts in various levels including technical level, the sector inter-state policy level (in ministerial forums such as the African Ministerial Conference on Environment [AMCEN], the African Ministers' Council on Water [AMCOW] and others dealing with energy, agriculture, housing and urbanisation, disaster reduction, meteorology among others) and at the continental and global summits (*ibid.*). At the regional level Zimbabwe is party to processes, instruments and agreements in the context of the Southern Africa Development Community (SADC) and the Common Market for Eastern and Southern Africa (COMESA). Locally, it has enacted several laws and bye-laws and set up institution for climate change response for example the Climate Change

The Zimbabwean government in efforts to localise global treaties it has enacted a number of laws and set up several institutions charged with the mandate to administer the respective environmental laws. The general observation is that, Zimbabwe, in terms of legislative and regulatory framework is better positioned for attainment of the global environmental treaties. Much work is needed to expedite and promote transparency and earnest implementation or crafting of support programs and activities for environmental sustainability. The major worry is political patronage wherein the policies or legislation amendment is done with political intends rather than genuine commitment to make cities and human settlements sustainable. The laws cover the whole spectrum of environmental facets including, flora and fauna, air, water and land management to protect the same from degradation resulting from human activities. Taking air quality as an example, Zimbabwe has the Atmospheric Pollution Prevention Act of 1971 (Atmospheric Pollution Prevention Act, 1971) (Phiri, 2009; Nickerson, 1994). This Act provides for the control of air pollution caused by noxious and offensive gases, smoke, dust and fumes from internal combustion engines. Zimbabwe's Ministry of Health is the agency responsible for administration of the Act (Melnick, 2005). The Act, seeks to implement the global air standards embracing the local context. The Air Pollution Advisory Board advises the Health Minister concerning the control, abatement and prevention of air pollution.

For water resources, Zimbabwe is mandated to protect the local as well as transboundary water resources. Prior to independence, Zimbabwe's primary legislation protecting water was the Fish Conservation Act. Although the Act's main purpose was the conservation of indigenous fish, it also contained a prohibition against water pollution. Specifically, the Act forbade the deposit or discharge into water of any substance that was injurious or potentially injurious to fish or to the marine life that sustained them. First offenses of this provision were punishable by fine or imprisonment for up to twelve months. The Act is no longer in force in Zimbabwe, however, because it has been repealed and subsumed by the Parks and Wildlife Act. Currently, Zimbabwe has three acts that relate to water resources: the Regional Water Authority Act (1976), the Water Act (1976) and the Zambezi River Authority Act (1987) (Mbula, 2016). Only the Water Act contains provisions of environmental significance, while all three of the acts are aimed primarily at balancing existing legal water rights and establishing commissions to regulate dam operation, water supply and usage and navigation upon the waterways. Zimbabwe protects its natural resources through a number of comprehensive laws (Cho *et al.* 2012). First, the Natural Resources Act provides for the conservation of a wide range of resources. Second, the Forest Act and the Mines and Minerals Act provide for the protection and management of these respective resources. The Natural Resources Act broadly defines natural resources to include soils, waters, minerals, animals, trees and vegetation, marshes, swamps and anything else the President proclaims to be a resource. The Ministry of Environment, Water and Climate is responsible for administration of the Natural Resources Act.

The 2013 constitution of Zimbabwe provides environmental rights in Section 73. This is also supported by section 4 of the Environmental Management Act (EMA) (Chapter 20:27), 2002, affords every citizen of Zimbabwe the following environmental rights:

- The right to live in a clean environment that is not harmful to their health; Access to environmental information;

- The right to protect the environment for the benefit of present and future generations; and
- The right to participate in the implementation of legislation and policies that prevent pollution, environmental degradation and sustainable management and use of natural resources, while promoting justifiable economic and social development.

The ministry with overall responsibility for the environment is the Ministry of Environment, Water and Climate. The general functions of the ministry is to regulate the management of the environment and to promote, co-ordinate and monitor the protection of the environment and the control of pollution; and regulate the activities of all government agencies and other agencies on their impact on the environment (Nhamo, 2002). It has the responsibility to administer the Environment Management (EMA) Act, 2002, [Chapter 20:27]. The Act provides a set of institutional set-ups and legal foundation for the sustainable management of natural resources and the protection of the environment; the prevention of pollution and environment degradation; the preparation of a national and other environmental management plans; as well as the establishment of an Environmental Management Agency and an Environment Fund (Mubvami, 2002). Part III of the EMA Act [Chapter 20:27] calls for the establishment of the National Environmental Council whose duties will be to advise on policy formulation and give directions on the implementation of the Act (Nhamo, 2002). The Council's major role is to recommend to all appropriate authorities' issues regarding the harmonization of functions related to environmental management. Also, the Council reviews and recommends incentives for the protection of the environment (Mubvami, 2002; Rajah *et al.* 2009). Part IV provides for the establishment of the Environment Management Agency (EMA) whose main duties is to advise the Minister on any matter pertaining to the planning, development, exploitation and management of the environment and in particular to develop guidelines for the preparation of a national plan and local environmental action plans.

The Agency also regulates; environment impact assessments; the management and utilisation of ecologically fragile ecosystems; and undertake, in the public interest, any works deemed necessary for the protection of the environment (Nhamo, 2003:16). For the purpose of promoting and facilitating the co-ordination of strategies relating to the environment, the Minister is mandated to prepare a National Environmental Plan as outlined in Part X (Nhamo, 2002; Mubvami, 2002). The National Environmental Plan shall formulate strategies and measures for the management, protection restoration and rehabilitation of the environment. In addition, all local authorities (Urban Councils, Rural District Councils and Town Boards) are required to prepare their own Environmental Action Plans. Most importantly, is the Environmental Impact Assessment (EIA) Requirements which have been stipulated under Part XI of the EMA Act (Mubvami, 2002). The EMA Act charges that all development projects with, especially significant negative environmental impacts shall be subjected to full EIAs. The EIA provides a detailed description of the project and the activities to be undertaken as well as the likely positive and negative impacts the project may have on the environment (*ibid.*). It also has to specify the measures proposed for minimising and where possible eliminate adverse effects and enhance positive ones. When the EIA report meets the set requirements, a certificate, valid for a period of two years, shall be issued. Under Part XII of the EMA Act, the President is empowered to set aside state or communal land for environmental purposes and with the help of the Minister declare it as a wetland or take measures necessary for the conservation of biological diversity (Nhamo, 2002:18). Part XIII empowers the inspectors to enter any land for the purpose of ascertaining if any invasive alien species are growing there (*ibid.*).

Zimbabwe's Environmental Actions in relation to the International Environmental Laws

Although Zimbabwe is party to numerous international environmental agreements, it should not be assumed that the country adopts entirely the prevailing, usually, Western-oriented view of environmental issues (Phiri, 2009). In fact, Zimbabwe, like many Third World nations, has voiced serious complaints about the degree to which international environmental law has been the product of primarily developed nations (Slaughter, 2009). In cooperation with other African states, Zimbabwe has sponsored several key initiatives aimed at establishing international environmental guidelines that better reflect the perspectives of African and other Third World nations. Whereas Zimbabwe supports the sustainable development concept in principal, internal realities have affected practical application of this concept. Primarily because Zimbabwe is in the unique circumstance of possessing many pristine nature reserves and environmentally sound regions that are yet to be developed, the country's leaders believe that Zimbabwe should be exempt from certain internationally imposed environmental restrictions (Cornwall, 2004). For example, while elephants are an internationally protected species under CITES, Zimbabwe actually suffers from an excess population of elephants. The Zimbabwean government therefore believes that existing international controls should not apply to its particular circumstances.

In the past, permits for new projects were granted in terms of the Regional, Town and Country Planning Act (Chapter 29:12). Under the new EIA Regulations (SI No. 7 of 2007), this has changed, and local authorities can only issue licences to developers after first seeing the licence from the Agency confirming that an EIA has been approved (Murombo, 2012; Mbida *et al.* 2006). In terms of the Regional, Town and Country Planning Act, the Ministry of Environment and Natural Resources Management is regarded as the local authority for parks, wildlife and forest lands. The developer must undertake an EIA for any developments in these specific land use areas (Chikaura, 2013). The Mines and Minerals Act regulates mining projects and requires an EIA to be undertaken and the policy condition to be met for these projects. Permits relating to water abstraction and water storage are granted in accordance with the Water Act. The Waste Disposal Licence is issued by the Agency and is renewable on an annual basis. The licence holder is obliged to pay inspection fees and environmental fees to the Agency. A licence is not required for household or domestic waste disposal or for the application of inorganic fertilisers for agricultural production.

In Zimbabwe, all projects, as prescribed in a schedule to the Environmental Management Act, must undergo EIA before implementation (Kalima, 2002). Compilation of the EIA report must be done by an independent consultant at the expense of the proponent and the basic contents of the document are guided by regulation. As part of the EIA report, an EMP (Environmental Management Plan) must be submitted to guide monitoring of implementation and operation of the project (El-Fadl and El-Fadel, 2004). The document is then assessed by EMA officers, with site visits conducted to verify suitability of proposed site and also stakeholder consultation (affected communities, businesses) carried out. After certification, bi-annual audits for compliance are carried out. The EIA program continues to evolve into an integrated network of program that work in support of an ecosystem approach to management, as well as develop Environmental Operating Guidelines for the various sectors. Table 10.1 as follows indicates the summary of globalising international environmental law into Zimbabwe through examining and comparing on how it has been applied in Zimbabwean context, its significance and trend of analysis.

Table 10.1: A Summary of Globalising International Environmental Law into Zimbabwe

Dispensation	Global Context	National Context
1960s-1970s	• The Convention for the Protection of World Cultural and Natural Heritage (1972) • The Convention on International Trade in Endangered Species (1973)	• Zimbabwe has also ratified the Convention for the Protection of World Cultural and Natural Heritage (1972) • The Zimbabwean government's efforts to protect air quality are channelled primarily through the Atmospheric Pollution Prevention Act of 1971. • After the CITIES, primarily because Zimbabwe is in the unique circumstance of possessing many pristine nature reserves and environmentally sound regions that are yet to be developed, the country's leaders believe that Zimbabwe should be exempt from certain internationally imposed environmental restrictions. • Parks and Wildlife Act (1975, Chapter 20:14)- The Act establishes national parks, botanical reserves and gardens, sanctuaries, safari areas and recreational parks provides for the conservation and control of wildlife, fish and plants; and designates specially protected animals and indigenous plants.
1980s	• The Law of the Sea Convention (1982);	• Bonn and Gaborone Amendments; the Vienna Convention for the Protection of the Ozone Layer

	• The Vienna Conventions on the Early Notification of and Assistance in the Case of Nuclear Accident or Radiological Emergency (1986) • Bonn and Gaborone Amendments; the Vienna Convention for the Protection of the Ozone Layer (1985) • the Montreal Protocol on Substances that Deplete the Ozone (1987) • the Brutland Commission of 1987 for	(1985) • With respect to regional agreements, Zimbabwe has participated in the First African Ministerial Conference on the Environment (1985), UNEP, supra note 60; • The Harare Action Plan for Environmentally Sound Management of the Zambezi River System (1987) • After the First African Ministerial Conference on the Environment, which was held in Cairo in 1985, was further reflected in Zimbabwe's delegation of responsibility for protecting the environment and developing tourism to a single ministerial department, indicating its view that economic prosperity and environmental conservation are inextricably intertwined. • The National Conservation Strategy of 1987 was the first policy document to incorporate the concept of sustainability into development and environmental management • Communal Land Forest Produce Act (1988, Chapter 19:04) - The Act controls the use of wood resources within communal lands. Such resources in communal lands should be used for domestic purposes by the residents only. • Rural District Councils Act (1989, Chapter 29:13)- The Act allows for the establishment of rural district councils responsible for initiating and regulating development in rural areas.
1990s	• the Agreement for the Establishment of Southern African Centre for Ivory Marketing (1991) • the Rio Convention on	• The environmental impact assessment policy published in 1997 by the then Ministry of Environment and Tourism (MET) is the current policy governing EIAs. • The 1997 National Environment Policy is therefore used in conjunction with the Environmental

	Biological Diversity (1992) • The Framework Convention on Climate Change (1992).	Management Act, the new EIA Regulations (SI No 7 of 2007) and the EIA Guidelines published by the ministry to ensure that EIAs are carried out correctly • Water Act, No. 31 of 1998- The Act regulates the planning and development of water resources and provides a framework for allocating water permits.
2000s	• The 2002 Johannesburg Summit	• In 2002, the Environment Management Agency was formed and section 4 of the Environmental Management Act (EMA) (Chapter 20:27), 2002, affords every citizen of Zimbabwe the environmental rights. • The Water (Waste and Effluent Disposal) Regulations of 2000, which are associated with the Water Act, No. 31 of 1998, specify what quality is acceptable in terms of effluent released into rivers. • In 2003, the new National Environmental Policy drawn up after the promulgation of the original Act is still in a second draft form and is not yet in effect. • The new EIA Regulations (SI No 7 of 2007) were also created in the 2000s. • the EIA and Ecosystems protection Statutory Instrument (SI) 7 of 2007, Solid waste management SI 6/2007; SI 98/2010, Atmospheric pollution control SI 72/2009, Hazardous substances and waste management SI 12/2007; SI 10/2007; SI 77/2007 AND Effluent Disposal control SI 6/2007.
2010-2016	• The Paris Agreement	• 2013 constitution of Zimbabwe provides environmental rights in section seventy-three which states that every Zimbabwe has a right to stay in a clean environment and environmentally sustainable.

It is clear that Zimbabwe, in coordination with other African states, is in the process of evolving its own international guidelines that it believes address the environmental realities of Southern Africa more effectively than existing regulations and procedures. Zimbabweans and other Africans understand that it is in their best interest to protect the environment that sustains them. Accordingly, Zimbabwe's bottom-up management approach to solving environmental problems, such as wildlife conservation, have been encouraged and though concerns are there peddling that this is hindered by the global community especially in wildlife managements

Gaps in Environmental Management in Zimbabwe

Despite several efforts by the Zimbabwean authorities to localise the international environmental laws through various legislations and policies, Zimbabwe is still in serious environmental problems emanating from continued shortfalls in its interaction with humankind. EMA has identified deforestation, drought and desertification, soil erosion and fires, water pollution, loss of biodiversity, water hyacinth invasion on lakes and dams, air pollution and poor waste management as some of the environmental challenges facing Zimbabwe. Whilst EMA is punitive to some degree to curb such activities mentioned above, it lacks adequate human and financial resources to enforce it. The environmental efforts have been largely derailed by lack of financial resources which has compounded the ability of authorities to acquire the needed materials, equipment, infrastructure and human resources for the task. This has compounded the ability of the authorities responsible for environmental law enforcement to effectively and appropriately administer the implementation of the statutes. Regional and local authorities do not have adequate resources or capacity to pursue integrated environmental management. Also, the worsening economic fortunes in the country has diverted efforts of many people and led to serous compromises in environmental law enforcement to allow people survival. This has seen a rise in environmental degradation, for example, the informal sector has not acted responsibly with the waste it generates and it discharges this waste into the environment. Shortage of housing in urban areas has also led to serious deforestation in peri-urban areas as people are turning these former farms or forests into housing; urban sprawling.

There is also lack of transparency of the law in Zimbabwe. Although the law is punitive, there is lack of transparency in the enforcement of such laws hence many large companies continue to pollute the environment. This lack of transparency is bred by corruption tendencies which seem to top on news in Zimbabwe. Officials may receive bribes from the culprits of environmental degradation with the effect of increasing the act, environmental degradation. Also, there is lack of effective stakeholder participation in environmental management. While stakeholders and third parties are integrally involved in the resources' management law review process, meaningful effective stakeholder participation is not present apart from a few exemplary cases. The capacity, the conditions and willingness to explore new participatory approaches are all missing in the process of localising global environmental efforts. More so, there is also lack of political will among government departments, parastatals, industry, farming community and the public in general to act responsibly or adhere to environmental laws. Examples include the reluctance by urban residents to segregate their waste when disposing it into bins. The authorities have put initiatives to separate waste for easy waste management, but to date, people are still mixing this waste; this is an activity that do not need any extra money or effort but just the will by the concerned person to adhere, but many are not doing so. The EMA Act does not define roles and responsibilities of sectorial ministries making the enforcement of environmental process overly bureaucratic for example the ownership and issuance of certain resources like water permits and allocation of mining rights are carried out under entirely separate laws. This creates bureaucracy in managing these resources under the EMA Act (Naome *et al.* 2012).

Conclusion and Policy Directions

Overall, Zimbabwe just like most African states has managed to establish laws and formal governmental structures to address the environmental problems as stipulated in international environmental laws, but very few have been successful in alleviating those problems. The environmental laws are impressive but lack implementation. In this light, there is need for an appropriate environmental management delivery strategy in form of integrated environmental management with enough institutional, financial, technological and human capital support. The government must also seek to foster stakeholder participation in environmental issues as this creates public buy-in; public buy-in is important as it reduces the efforts needed on enforcement as people will be policing themselves and also it may generate resources in form of local knowledge, human or financial resources which are lacking at the moment. Participation of stakeholders also assists in information dissemination which is crucial in the creation of environmental stewards. The government is also recommended to deal decisively on corruption and abuse of office, this has costed the nation much resources which can be channelled towards project of environmental significance and also led to arbitrary ignorance of the law; people are ignoring the environmental laws constructing on wetlands, discharging waste inappropriately on undesignated site among other environmental ills.

References

Arts, B., and Buizer, M. (2009). Forests, Discourses, Institutions: A discursive-institutional analysis of global forest governance. *Forest policy and economics,* 11(5), 340-347.

Bekele, M. (2008). Ethiopia's environmental policies, strategies and programs. Digest of Ethiopia's national policies, strategies and programs. FSS, Addis Ababa, Ethiopia, 337-69.

Biswas, A. K. and Agarwal, S. B. C. (Eds.). (2013). *Environmental impact assessment for developing countries.* Amsterdam, Elsevier.

Brunch, C., Coker, W., and VanArsdale, C. (2001). Constitutional environmental law: Giving force to fundamental principles in Africa. *Colum. J. Envtl. L., 26,* 131.

Chapman, K., and Walmsley, B. (2003). Country Chapter: Zambia. In: SAIEA (Southern African Institute for Environmental Assessment), EIA in southern Africa. Windhoek: SAIEA.

Chikaura, F., and Frank, B. (2013). Stakeholder Analysis Report. Available online: https://www.birdlife.org/sites/default/files/attachments/blz_stakeholder_capacity_building_assessment_report.pdf. [Accessed on 16 June 2018]

Chinamhora, W., and Ruhukwa, D. (1995). Towards an Environment Management Act Review and Revision of Zimbabwe's Environmental Legislation. Harare, Ministry of Environment and Tourism.

Cho, M. E., and Tifuh, J. (2012). Quantification of the impacts of water hyacinth on riparian communities in Cameroon and assessment of an appropriate method of control: the case of the Wouri River Basin. Available online: https://commons.wmu.se/cgi/viewcontent.cgi?article=1028&context=all_dissertations, [Accessed on 16 June 2018]

Cornwall, A., and Nyamu-Musembi, C. (2004). Putting the 'rights-based approach' to development into perspective. *Third world quarterly*, 25(8), 1415-1437.

Damtie, M., and Bayou, M. (2008). Overview of Environmental Impact Assessment in Ethiopia. *MELCA Mahiber*, Addis Ababa, Ethiopia.

El-Fadl, K., and El-Fadel, M. (2004). Comparative assessment of EIA systems in MENA countries: challenges and prospects. *Environmental impact assessment review*, 24(6), 553-593.

Environmental Law Institute. (2007). *Constitutional environmental law: Giving force to fundamental principles in Africa*. Washington, D.C. Environmental Law Institute.

EMA. (2002). Environmental Management Act Chapter 20:27 Harare, Government Printers

Glasson, J., Therivel, R., and Chadwick, A. (2013). *Introduction to Environmental Impact Assessment*. London, Routledge.

Government of the Republic of Zimbabwe. (2013). Constitution of the Republic of Zimbabwe. Harare, Government Printers

Kalima, J. (2002). Environmental Impact Assessment as a Tool of Common Property Resource Management in Southern Africa: Towards an Integrated Regime. IASCP 9th Biennial Conference.

Mbiba, B., and Ndubiwa, M. (2006). Decent work in construction and the role of local authorities the case of Bulawayo City, Zimbabwe.

Mbula, M. (2016). Impacts of Water Hyacinth on Socio-Economic Activities on Kafubu River in the Copperbelt Province.

Melnick, D., McNeely, J., and Navarro, Y. K. (2005). *Environment and Human Well-Being: A Practical Strategy*. London, Earthscan.

Mohamed-Katerere, J. (2001). Participatory natural resource management in the communal lands of Zimbabwe: What role for customary law. *African Studies Quarterly*, 5(3), 1-27.

Morgan, R. K. (2012). Environmental Impact Assessment: The State Of The Art. *Impact Assessment and Project Appraisal*, 30(1), 5-14.

Moyo, S., Robison, P., Katerere, Y., Stevenson, S. and Gumbo, D. (1991). Zimbabwe's Environmental Dilemma Balancing Resource Inequalities. ZERO, Harare.

Moyo, S., Sill, M., and O'Keefe, P. (2014). *The southern African environment: Profiles of the SADC countries.* London, Routledge.

Mubvami. T. (2002). Environmental Impact Assessment as a Policy Tool for Environmental Management. In IUCN-ROSA. (EDT), Approaches to Environmental Policy Analysis: Southern Africa. Harare, Zimbabwe: IUCN-ROSA.

Munowenyu, E. (1999). Introduction to Geographical Thought and Environmental Studies., Harare, Zimbabwe Open University.

Murombo, T. (2012). Balancing interests through framework environmental legislation in Zimbabwe. 2012, 557.

Nhamo, G. (2002). Institutional and Legal Provisions for Environmental Management in Zimbabwe. *Ajeam*-Ragee, 1(7), 14-20.

Nicholas A. R. (1994). Agenda 21: Working towards a Global Partnership, men
Environmental Policy and Law Paper, No. 27 (1994). Available online:
http://researchbriefings.files.parliament.uk/documents/RP96-87/RP96-87.pdf. [Accessed on 15 June 2018]

Nickerson, B. J. (1994). The Environmental Laws of Zimbabwe: A unique approach to management of the environment. BC Third World LJ, 14, 189. Available online:
http://lawdigitalcommons.bc.edu/twlj/vol14/iss2/1/ [Accessed on 15 June 2018]

Norval, R. A. I., Walker, J. B., and Colborne, J. (1982). The ecology of *Rhipicephalus zambeziensis* and *Rhipicephalus appendiculatus (Acarina, Ixodidae)* with particular reference to Zimbabwe. Available online:
https://repository.up.ac.za/bitstream/handle/2263/51657/35n orval1982.pdf?sequence=1. [Accessed on 15 June 2018]

Parks and Wildlife Act (Chapter 20:14). (1996). Revised Edition, Harare, Government Printers.

Phiri, S. (2009). Impact of artisanal small-scale gold mining in Umzingwane District (Zimbabwe), a potential for ecological disaster. Available online:
https://www.researchgate.net/publication/306235182._[Accessed on 15 June 2018]

Rajah, N., Rajah, D., and Jerie, S. (2012). Challenges in Implementing and Integrated Environmental Management Approach in Zimbabwe. *Journal of Emerging Trends in Economics and management Science*, 3(4), 408-414.

Sadler, B., and Dalal-Clayton, D. B. (2012). Strategic environmental assessment: a sourcebook and reference guide to international experience. Earthscan.

Shinn, D. H. (2016). The Environmental Impact of China's Investment in Africa. Cornell Int'l LJ, 49, 25.

Slaughter, A. M. (2009). A new world order. Princeton University Press.

Sterner, T. (2003). Policy instruments for environmental and natural resource management. Resources for the Future.

Sunderlin, W. D., Angelsen, A., Belcher, B., Burgers, P., Nasi, R., Santoso, L., and Wunder, S. (2005). Livelihoods, forests and conservation in developing countries: an overview. *World development*, 33(9), 1383-1402.

The Constitution of the Republic of Zambia (as amended by Act No. 18 of 1996). The Environmental Management Act (EMA), No. 12 of 2011.

The Hazardous Substances and Articles Act (Chapter 15:05). Revised Edition (1996), Harare, Government Printers.

Water Act. (1998). Water Act (Chapter 20:24), Harare, Government Printers.

Wunder, S., Engel, S., and Pagiola, S. (2008). Taking stock: A comparative analysis of payments for environmental services programs in developed and developing countries. *Ecological economics*, 65(4), 834-852.

Zhang, Q., and Crooks, R. (2012). Toward An Environmentally Sustainable Future: Country Environmental analysis of the People's Republic of China. Asian Development Bank.

Chapter 11

The Food-Water-Health-Energy-Climate Change Nexus: Pivot for Resilience in the Cities of the Global South

Innocent Chirisa, Verna Nel and Romeo Dipura

Introduction

Global climate change and resource constraints have emerged as some of the major challenges in the 21st century (Carter and Gulati, 2014). The long-term effects of climate change which include changes in rainfall patterns, rising temperatures, increase in the frequency of climate related shocks negatively affect health, food, energy and water supply (Wheeler *et al.* 2000). The interconnections and interdependence between food, energy and water resources amplify the negative impacts of climate change. Although climate change is a global phenomenon, countries differ in terms of exposure, vulnerability and their capacity to adapt to climate change. Furthermore, the effects of climate change are experienced differently within countries, with positive effects seen in some regions and negative effects in others (Hoff, 2011). The countries in the Global South have been most vulnerable to the negative effects of climate change. The issues of energy, food, and health and water security have also risen to global prominence as they affect increasing numbers of people in an interconnected world. All societies rely depend on energy, food, health and water to survive. However, there are hundreds of millions of people who are lacking access to these basic necessities in sufficient quantities and of adequate quality (Cilliers, 2008).

Food, water, health and energy security form the backbone of all self-sufficient economies (Tadesse, 2010; Cilliers, 2008; IPCC, 2014). Shocks in any one of these systems directly affect the others (Sachs, 2010). It is particularly important to note the effects of climate change on food security. As developing countries experiences water-scarcity with little arable land and thus, increasingly dependent on oil imports as economies in the Global South test the limits of resource constraints (Carter and Gulati,

2014). The ability of developing countries to protect their food security from the negative impacts of climate change depends on the extent to which the risks and the vulnerability of food security to climate change are understood especially from the perspective of the nexus between food, water, energy and health. The energy-food-water nexus has thus emerged as an important focus within international development, sustainability and policy discourses. In a continuously resource-constrained world, the nexus between food, water, health and energy which can be defined as the interconnections among these three systems that is vital for human survival (UNEP, 2014).

The health-energy-food-water nexus comprises a multidimensional, complex set of issues operating at various levels from the global to the national scale. The nexus manifests differently in urban areas as compared to rural environments, partly due to the various components of energy, food and water systems. The health-energy-food-water nexus is by nature a complex and interconnected set of issues, which can be approached in a variety of ways (IPCC, 2014). Health, water, energy and food are important for human wellbeing and sustainable development. These four systems are intricately linked as food production needs water and energy; water management demand energy (Foresight, 2011). Hydro-energy production on the other hand which is the major source of energy in southern Africa also requires water. Any impact on one affects the others. The interconnections between these four systems are referred to as the health-food-energy-water nexus which when adequately understood can perpetuate socioeconomic securities and development (UNDP, 2007). This four-way interaction between these systems has resulted in the emergence of the health-food-energy-water nexus as an important concept for integrated resource management at various spatial scales.

This chapter explores the nexus between water, food, health, energy and climate change systems and examining the possibility of capitalising on synergies in answering the development challenges in the Global South. It explores the complex relationship between food, water and energy systems from the perspective of the sustainability agenda. It is focused on reviewing the integrated system that supplies energy, water and food for use within the countries of the Global South.

Theoretical Framework

This section is focused on examining the various theories relating to the health-energy-food-water nexus. The concepts that are explored are climate change, health-energy-food-water nexus, vulnerability, risks and resilience. The health-energy-food-water nexus is emerging as an important discourse of sustainable development in this increasingly resource constrained world. Carter and Gulati (2014) define the health-energy-food-water nexus as interconnections between health, energy, food and water systems which comprises of multidimensional, complex set of issues operating at various scales from the global to national to local levels. The nexus also manifests itself differently between rural and urban environments (Wakeford and de Wit, 2013).

Figure 11.1 illustrates the interconnections that exist between food, water and energy and health system in an environment that is liable to the impacts of climate change. Some of the major impacts of climate change are; increasing average temperatures, changes in the patterns and quantity of rainfall, increased severity of droughts, increased intensity of extreme events and this has had both direct and indirect repercussions (World Bank, 2013). The most direct impact climate change is expected to have on food security is through changes in crop and livestock productivity. Climate change variability immensely affects availability, accessibility and affordability of food (UNDP, 2008). It affects the stability of the food system directly through changes in productivity, quality of yield, crop failures, loss of livestock, farming costs and the effects of changing weather conditions on agricultural practices. Impacts of climate change on the food system can also be indirect by potentially affecting water resources and the distribution of pests and diseases (UNEP, 2013). Climate change has also induced more frequent and intense weather events which include rising sea levels resulting in irregularities in seasonal rainfall patterns which are negatively affecting both food production and distribution infrastructure (FAO 2008).

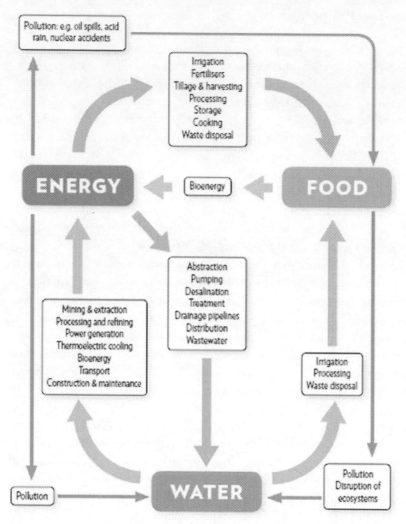

Figure 11.1: Conceptual framework for health-water-food-energy nexus (IRENA, 2015)

Decreased rainfall for long periods has resulted in the recurrence of droughts, with more frequent high-intensity rain events exacerbating soil degradation and desertification (Hoff, 2011). Soil erosion has been creating problems in the viability of agriculture, as nutrient depletion facilitates crop failure as more quantities and types of inputs would be required in enhancing soil productivity (United Nations, 2013). However, climate change variability and change could also benefit food production, although the gains and losses will vary across farming systems and provinces.

Furthermore, higher temperatures and humidity reduce agricultural yields as well as encourage the proliferation of weeds and pests as high carbon-dioxide concentrations favour weeds more than agricultural crops (Foresight, 2011). Ultimately climate change negatively affect agricultural production, prices and infrastructure will change thus limiting the amount and quality of food produced (Wlokas, 2008). Rising temperatures and changes in rainfall patterns have a direct effect on crop yields, as well as an indirect effect through changes in the availability of irrigation water.

The WEF (2013) describes water as a commodity without any substitute which connects humans, the environment and, all aspects of the economic system. Water is critical for the growth of biomass thus it is a requirement for all ecosystem services on which the livelihoods and economic systems depend (Hoff, 2011). A large share of global water sources is used in agricultural activities as agriculture accounts for 70% of the global freshwater withdrawals. It is also worth noting that water resources also contribute to energy systems with energy accounting for 20% of the global freshwater withdrawal (UN-Water, 2014). Climate change negatively affect water supply as rising temperatures induce higher evapotranspiration as well as reducing run off. Changes to the frequency and intensity of rainfall lead to the increased incidence of droughts and floods. Water supply also affect the energy system as adverse changes in the quality and quantity of water resources require increased energy inputs for water purification (UNEP, 2013). Lower quality water or pumping water from large distances intensifies the competition existing between the energy and food sectors for the existing water resources. Other impacts of water supply on energy system may be in the context of renewable energy (wind speeds, solar radiation levels and hydro-energy).

At the same time, climate policies may have implications for the health-water-food-energy nexus. For example, the use of carbon sequestration, expansion of biofuels or hydropower as measures of mitigating the negative impacts of climate change can create new water demands (IPCC, 2014). Climate change adaptation can usually be energy intensive as the case is with irrigation systems; rain fed agriculture as well as desalinisation (OECD, 2012). There are higher energy demands for pumping with increased use of underground water. The adoption of risk management measures and adaptation strategies to enhance resilience against the negative effects of climate change alter the potential impacts of climate change (Hoff,

2011). More so, the increasing incidence of water-borne diseases in areas prone to flooding, changes in vectors for climate-responsive pests and diseases, the emergence of new diseases affect the food chain as well as the human capacity to obtain nutrients from the foods they consume (Tadesse, 2010).

Energy systems are connected to food and water systems through dependencies on them as inputs and through externalities (World Bank, 2013). Changes in cloud cover, wind patterns and rainfall which can be induced by climate change negatively impact renewable energy production as hydropower is chiefly vulnerable to a drier climate. Climate change can also indirectly increase the amount of energy required by the country through adaption policies. Increasing irrigation is a response to reduced water supplies as rainfall decreases as a result of climate change (Fargione *et al*. 2009). Higher temperatures also induce high demand for air conditioning. Biofuels on the other hand, require water to grow and also take arable land away from food production. The energy system also requires water inputs at various stages of the energy production and consumption chain (Rodriguez *et al*. 2013). Initially, water is used in the primary extraction of fossil-fuel mining and production.

Large volumes of water are also required in the extraction of unconventional resources such as in the production of synthetic oil from tar sands, for enhanced oil recovery as water is injected into conventional oil wells as well as in the hydraulic fracturing processes used to extract oil and gas from shale basins. Water is also required to produce bioenergy crops such as sugar cane, soybeans and maize (UNEP, 2013). Other biofuels use green water which accumulates in soils as opposed to blue water which is provided for through infrastructure. Large volumes of water are also required in shifting energy from one form to the other such as in the washing of coal to prepare it for use in power stations. Hydroelectricity directly depends on water supply and some forms of geothermal power are transmitted through underground water. Water is also a key in the construction of energy infrastructure such as refineries, power grids, power plants and pipelines (UNDP, 2014). The operation and maintenance of energy infrastructure also requires water. Conveyance of raw water often requires energy for pumps. Further energy is required for water treatment and purification. These energy requirements vary, depending on the quality of the input water, the type of treatment technique and the use to which the water will be put. Drinking water requires extensive treatment, while irrigation water requires little or no treatment. Distribution of water through pipeline systems to consumers depends on reliable supplies of energy, usually in the form of electricity.

Water systems to a great extent also depend heavily on energy at many stages along the water supply chain (Rodriguez *et al.* 2013). Energy in the form of electricity or diesel is required in the extraction phase is needed to pump water from lakes, rivers and underground aquifers. The amount of energy required to pump water depends on factors such as distance, elevation change, pipe diameter and friction determine the amount of energy required (UN-Water 2014). Extracting groundwater is typically more energy-intensive than extracting surface water (Hoff 2011). Processes such as desalination of seawater are highly energy intensive and are mostly done in water-scarce and energy-rich regions such as the Middle East and North Africa. Extraction fossil fuels on the other hand pose threats to water quality. Mine tailings resulting from surface coal mining for example, they contain pollutants that can leach into groundwater, causing acid mine drainage (IRENA 2015). Oil and gas extraction results in large volumes of associated 'produced water', which is often costly and difficult to treat.

The direct dependence of the energy system on the food system relates to the conversion of some food crops into bioenergy, particularly liquid biofuels (FAO, 2014). An additional way in which energy systems may depend on food systems is through the use of food waste to generate bioenergy such as methane gas (biogas); however, the amounts involved even at a global scale are negligible. Consistently, to note in the nexus context is the conversion of food crops into modern biofuels for transport, electricity and heating (UNEP, 2014). Negative externalities emanating from energy systems and impacting on food systems mostly relate to contamination of water resources. However, pollution from extractive activities can also have effects on soil fertility. Energy can also indirectly affect agricultural production through the contribution of fossil-fuel combustion to greenhouse- gas emissions and climate change. New biofuel plantations have the potential to carry a negative carbon balance for many years (Fargione *et al.* 2009). Furthermore, bioenergy production also negatively affects agricultural production. Likewise, to note is the impact of bioenergy production on agricultural output, via the competition for limited land and water resources (UNDP, 2007). The food system depends on energy at all stages of the value chain from inputs for production and processing right through to consumption and waste management (WEF, 2011). The specific linkages and dependencies can be analysed in each stage of the value chain. The food system is broader than just agricultural production. The global food system comprises entire value chains for each crop, livestock or fish product, from inputs to waste.

According to Gitz and Meybeck (2012), the concepts of risk, resilience and vulnerability are interconnected. These concepts interact in their application to biophysical, social and economic processes. Risk refers to the existence of potential shocks which can affect the state of systems, households, communities and individuals. Vulnerability on the other hand refers to the propensity or predisposition to be negatively affected by a risk. The concept of risk is very dynamic varying across space and across time and depends on environmental, cultural and social factors. Vulnerability is a complex concept as it demands consideration across diverse dimensions. When considering vulnerability and risk it is also essential to consider resilience and adaptive capacity. Resilience refers to a system's ability to anticipate, accommodate as well as absorb the negative effects of a hazardous event timeously and in an efficient manner by repairing, preserving or restoring the basic structures of the system. Adaptive capacity refers to a system's capacity to adapt such that it is less vulnerable. Smit and Wandel (2006) identify two aspects to adaptive capacity namely coping ability and adaptability which is mainly related to time.

Literature Review

The world is increasingly becoming resource-constrained and the health-energy-water-food nexus has become a panacea for sustainable development in countries of the Global South (United Nations, 2013). Furthermore, agriculture, food and nutrition security and natural resources form part of the focus areas identified to contribute to the ultimate eradication of poverty as indicated in the SADC Regional Poverty Reduction Framework (Carter and Gulati, 2014). The health-energy-food-water 'nexus' is defined as the interconnections between health-energy, food and water systems. It comprises a multidimensional, complex set of issues operating at various scales from the global to national to local scale. The nexus manifests differently in urban environments as opposed to rural environments (World Bank, 2013). The energy-food-water nexus is by nature a complex and interconnected set of issues, which can be approached in a variety of ways. These interactions are dynamic and non-linear and can encompass positive and negative feedback loops, thresholds and tipping points.

It is also worth noting that these interactions give rise to the emergent properties of the system; the properties are not contained within the individual components themselves (Cilliers, 2008). This principle is recognised in this nexus research by focusing on the interactions among the health, energy, food and water systems, each of which is complex in its own right. Land is a central issue in the nexus and in particular the various uses to which it can be put affect how the system is balanced (UNDP, 2008). Thus, a lack of coordinated health-water-energy-food nexus planning would threaten stability in each of the systems. Addressing food security in rural areas often relates to people's access to land, water as well as other productive inputs. Another key issue is the use of water for irrigation (UNEP, 2014). In expanding the area under irrigation, the region should consider adopting the health-energy-food-water nexus for integrated resource planning and management. This has the potential to remedy the current scenario where agriculture accounts for a large portion of the available water withdrawals, yet food insecurity remains a major challenge (Foresight, 2011). Although agriculture will always have a larger share of water withdrawals, improved agriculture water management will result in water savings and release more water to other sectors. From a strategic perspective, climate models estimate a decrease of about 20% in annual rainfall by 2080 in southern Africa (Carter and Gulati, 2014).

The health-energy-water-food nexus is subject to several major drivers. Demand-side drivers include population growth, economic growth, rising affluence, shifting consumption patterns, urbanisation and globalisation (Tadesse, 2010). Supply-side drivers include the depletion of conventional fossil fuel reserves and the degradation of soils, fresh water supplies and ecosystems. Climate change is anticipated to exert increasing pressure on water resources and destabilise agricultural production as well as various forms of energy generation (FAO, 2014).

The major nexus linkages are: energy inputs are required at all stages of the food system value chain, crop and livestock production, processing and storage, distribution, food preparation and disposal of food waste. Furthermore, agricultural crops are converted into bioenergy whereas water is critical for agricultural production, food processing and waste disposal (United Nations, 2013). Energy is also essential at many stages of the water system value chain, including abstraction, desalination, treatment, construction of storage infrastructure, pumping and waste-water treatment. Water is required for the extraction and processing of fossil fuels, generation of hydroelectricity and geothermal power, cooling within thermal power stations and production of bioenergy (HOFF, 2011). Certain energy industries and high-input agricultural production can also have adverse impacts on water and soil quality.

Treating health, energy, food and water systems independently of each other results in system linkages and vulnerabilities being underappreciated possibly leading to the formulation and implementation of ineffective policy measures (UNDP, 2007). It is thus important to understand the complex interactions occurring between health, energy, food and water systems so as to identify the major vulnerabilities and risks confronting developing countries. An understanding of the health-food-energy-water nexus is also important in informing planning and policy in developing countries as a means of mitigating these risks and promoting social equity, economic efficiency and environmental sustainability (UNEP, 2014). Another key concept relating to the health-food-energy-water nexus is the concept of metabolism of a society. This refers to the ways by which energy and materials are used satisfying human needs and wants, which is likened to the way water, food and minerals are processed in an individual human's physical metabolism (IPCC, 2014).

According to FAO (2014), food security refers to the physical, economic and social access to adequate, safe and nutritious food. Energy security on the other hand refers to the availability of energy sources to the citizens which is affordable and uninterrupted. Water security refers to the extent to which a population is capable of safeguarding access to sufficient quantities of water at the right quality which sustains livelihoods, wellbeing as well as socioeconomic development. Social and ecological systems interact through health, food, energy and water systems (Cilliers, 2008). Poor water quality negatively affects food and energy systems. Polluted water also has a detrimental impact on the quality of agricultural and food products. Water of poor quality also weakens some forms of operations of thermal power stations (Rodriguez *et al*. 2013).

According to Hoff (2011) to produce a calorie of food at least a litre of water is required. It is also important to note that water of poor quality has negative impacts on food production and some forms to energy generation. Water scarcity at a global scale is progressively being recognised as threatening human development. These challenges are also exacerbated by the negative impacts of climate change (FAO, 2007). In light of current and forecasted increases in global demand for health, energy, food and water resources, as well as the strong linkages between these systems managing the nexus between these systems is continuously being seen as a viable approach in managing sustainable development (UNEP, 2013). The health-food-energy-water nexus is thus a framework capturing the interconnections, synergies and trade-offs between these systems in the context of the emerging threats of climate change. The thinking behind the health-food-water-energy nexus is based on the systems approach with the socio-ecological system as the primary reference source. It has emerged as one of the major approaches for achieving (Goals 1, 2, 6 and 7) of the Sustainable Development Goals (SDGs) which are: poverty alleviation, zero hunger, provision of water and sanitation and access to affordable and reliable energy (UNDP, 2007).

The health-food-energy-water nexus can also play a prominent role in regional socioeconomic security, integration and security as deliberated in the WEF Nexus Action Plan, which is part of the Regional Strategic Action Plan IV (RSAP IV) (Carter and Gulati, 2014). Regions which face health, water, and food and energy challenges such as Southern Africa can immensely benefit from the health-food-energy-water nexus approach as it unlocks positive synergies which are pertinent in catalysing regional development (United Nations, 2013). These results can be brought about by ensuring coordination between countries through joint investments in energy projects whereas ensuring that water saved in one country can be freely released for agriculture in another country.

Health-food-water-energy nexus: Cuba case study

Cuba has had positive strides towards achieving food and energy self-sufficiency but however remain economically dependent on food and fuel imports. Fuel constitutes an average of 35% and food 15% of the total import bill in Cuba (Wakeford *et al*. 2015).

Although there is a minimal direct reliance of the energy sector on water, hydropower constitutes up to 0.6% of the country's electricity mix. Some energy is generated from bagasse and bioethanol thus the agricultural sector in Cuba contributes considerably to energy production. 4 138 mega tonnes of sugarcane bagasse, 1.2 million m3 wood and rice husks (16.5Mt) as well as waste from the coffee industry is utilised as feedstock for power generation. The success of the sugarcane sector in Cuba has however come at the peril of the country's forests, leading to massive deforestation during the 19th and early 20th centuries thus causing soil fertility loss and erosion. However, most of the sugar is produced for food rather than fuel and the energy component relies on agricultural waste, biomass energy has a limited negative effect. The energy sector however affects food output through competition for land and water. To note also are the negative impacts of the energy sector on the water and food sectors pertains to the pollution risks associated with oil extraction. Resource depletion and environmental degradation are the major factors in the food-energy nexus. Finite oil and gas reserves in Cuba are depleting and largely depend on imported oil, supplies of which may be constrained in the future (*ibid*).

South Africa Case study

In South Africa, coal is 69% of primary energy production is obtained from coal. Apart from coal as the main energy source, there is also oil which constitutes 15% of energy production followed by biomass which constitutes 10.7% (Wakeford *et al.* 2015). In South Africa, water is required at various phases in energy production, processing, transformation as well consumption and use cycle. The energy sector significantly relies on reliable water supplies although constituting only 2% of the abstracted water consumption. However due to the fact that hydro-power contributes insignificantly (2%) to power generation in South Africa, it has a relatively small water footprint of that technology. Water is also extensively required by the energy sector in South Africa for coal fired generation for scrubbing and cooling in Eskom power stations. The use of food products for bioenergy in South Africa is relatively low despite the fact that the sugar industry in South Africa has been using bagasse waste for years for electricity generation thus the energy sector doesn't rely too much on the agricultural system. Furthermore, biodiesel is produced from recycled vegetable oil and used in the transport market by small-scale operators without however negatively impacting on the food system; instead, it represents an efficient re-use of a waste product from the food industry although on a small scale. Energy production in South Africa has had negative impacts on water quality, as well as on agriculture and food production as it relies on high-quality water input. Coal mining in South Africa poses various risks on the water system which in turn negatively affect the agricultural system. The food system in South Africa also heavily depends on the energy system. Farm vehicles and machinery are powered by liquid petroleum fuels and irrigation systems rely heavily on electricity. The food system in South Africa is also extremely dependent on the water system. Only 12% of land in South Africa land is suitable for farming rain-fed crops. Although only 1.5% of South Africa's land area is under irrigation, it contributes 30% of crop volumes. 90% of horticultural production in South Africa and 12% of land planted to wheat are irrigated and the agriculture sector constitute up to 60% of freshwater withdrawals in South Africa (*ibid*).

Methodology

This chapter is based entirely on a desktop research which involved an extensive use of secondary data. It takes an exploratory approach on the health-food-energy-water nexus approach in building sustainable and resilient cities in the developing world. The study is to a greater extent qualitative in nature and involved an extensive use of secondary data. In line with its qualitative nature, the study takes an interpretivist research paradigm where reality is seen as an intersubjective construct. In this research paradigm meanings and understanding are products of social and experiential processes and knowledge is created by social and contextual understanding (Cresswell, 2003). The chapter also made use of a few case studies from which sought to expose the synergies between health, water, food and energy systems and the role of that is played by a nexus approach in advancing sustainable development in the cities of the Global South.

Results

The research findings revealed that climate change has significant impacts on the health-food-energy-water nexus in the countries of the Global South. The results also showed the various divers for the health-food-energy-water nexus and these are: population growth, urbanisation and global economic growth. The study also identified the health-food-water-energy nexus as a framework for sustainable resource planning.

Health-food-energy-water risks in the Global South

The results from the study revealed that whereas upper-middle-income countries (UMICs) are performing well as far as basic health, energy, food and water security is concerned access, most countries in the Global South (Africa, Asia, Latin America) have high levels of income inequality and poverty (UN Water, 2014). According to FAO (2007), globally, water scarcity affects 1.2 billion people in urban areas. Inadequate provision of water and sanitation services has also induced environmental and health challenges which have plagued cities of the Global South. Nostrom (2007) posits that the greatest proportion of the 5000 childhood deaths which occur daily on a global scale are due to diarrhoea which is linked to poor provision of water and sanitation services. Furthermore, oil price shocks are a significant global-scale risk within the energy system (Wakeford and DeWit 2013). Geopolitical events in oil-producing countries and regions such as the Middle East and North Africa, present risk in sudden rises in terms oil prices (Foresight, 2011). On the other hand, the risk of rising food prices threatens affordability of food for millions in the Global South especially for the poor in the developing countries. Shocks in food prices is usually a result of a growing world population, rising incomes perpetuating dietary changes as climate change as well as availability of arable land. The research findings revealed that 1.3 billion people are lacking access to electricity most of which are living in sub-Saharan Africa, Latin America as well as South and East Asia whereas a further 1.2 billion having unreliable access (World Bank, 2013). Over 780 million people also lack reliable access to potable water for drinking and sanitation (World Bank 2013). It is also estimated that 805 million people experience chronic undernourishment, constituting 13.5% of the combined population of developing countries (FAO, 2014). The most important point is not that energy resources, food and water are becoming economically scarcer as demand grows more rapidly than supply thus inducing higher prices but that the interconnections and interdependencies of these systems which are critical for human life are emerging as increasingly important (*ibid*).

Growing demand for water will place a major strain on agriculture in developing countries (IRENA, 2015). It is predicted that for irrigation withdrawals to meet growing food demand in the Global South by 2050 it has to increase at least by 11% by the year 2050 (Foresight, 2011). Most of the food demand will rise in countries that are already water scarce and have multiple competing demands on their water resources such as the countries in sub-Saharan Africa.). The research findings also revealed that although the notion of virtual water and increasing trade from water-rich countries has been advanced to assist water-scarce countries appears to be a viable approach, many factors limit this in practice (WEF, 2011). According to Chadwick (2012) volatile climate and the increased frequency and intensity of extreme weather events are among the top five factors shaping the citrus industry in the next 5 to 10 years in South Africa. The results also show that volatility in supply patterns, prices and the impact on quality are among the most important challenges that will be faced due to climate change. Climate change has also been noted to facilitate desertification and depletion of arable land in South Africa which presents serious challenges as agricultural land is limited with only 13% (UNEP, 2014). The research findings revealed that the major risks in relation to the health-water-food-energy nexus are extreme weather events, oil price shocks; food price shocks; geopolitical tensions; and financial speculation in commodity markets.

Climate change and the health-food-energy-water nexus

The negative impacts of climate change which induces changes in temperatures, rainfall patterns and the recurrence of extreme weather conditions negatively affect crop and livestock production as well as water and energy production. It is anticipated that rain fed agriculture will drop by at least 50 % in sub-Saharan Africa by 2020 (IPCC, 2014). The research findings also revealed that agricultural and food systems of the Global South also contribute to climate change. Agriculture in the Global South contributes to climate change through energy use, land-use change, methane emissions from livestock and irrigation practices in rice growing and nitrous

219

oxide emissions from fertiliser used on soils (Hoff, 2011). It is estimated that if current agricultural systems continue more than a billion hectares of virgin land will be converted to agricultural use to feed the global population by 2050 (Sachs, 2010). According to Foresight (2011), the largest contribution that the food system makes to climate change, followed by the production and application of nitrogen fertilisers and methane from livestock. However, the research findings also revealed that climate change can also create favourable climatic conditions for food production for a few locations. The negative impacts of climate change have had negative implications for health, energy, food and water systems of the Global South (IPCC, 2014). The results also show that the nexus linkages and feedback loops create a web which interconnect and reinforces risks and impacts. More so, the end result of these threats to health, food, and energy and water security is heightened social instability within countries of the Global South.

Drivers for the health-food-energy-water nexus

The results also show that the demand for health, energy, food and water will continuously grow in the next half-century due to a variety of factors (Sachs, 2010). These several systemic drivers operate at a global level in affecting all components of the nexus. On the demand side, these include: economic growth, increasing affluence and associated changes in lifestyles and consumption patterns; population growth and changing demographic profiles; urbanisation (OECD, 2012). On the other hand, supply side drivers include resource depletion, environmental degradation and climate change. Climate change is however both the result of processes in the health-energy-food-water nexus and a cause of instability and insecurity in some parts of the health, energy, food and water systems especially in water supplies and crop yields (IRENA, 2015).

The global population is projected to increase to 9.6 billion by 2050, with more than half of that growth occurring in Africa (United Nations, 2013). Moreover, the global economy is anticipated to multiply in size by mid-century, with rising living standards in developing countries leading to increased volumes and more resource-intensive patterns of consumption (OECD, 2012). Population growth will be one of the major drivers of change across all systems, including food, health, water and energy. This growth is projected to occur mainly in low-income countries; Africa's population is expected to double by 2050 (UNPD, 2008). The research findings also exposed the role played by urbanisation, especially in Africa, in raising resource demands as urban areas consume more resources as compared to rural areas (UNEP, 2013). The demand for energy is thus expected to increase by 80%, food by 60% and water by 55% (FAO 2014). Smaller cities in Asia and Africa are expected to grow substantially which further raises demand for health, food, water and energy services (UNPD, 2007). According to Hoff (2011) apart from the additional pressure exerted on resources such as water and energy more food will be needed. As food is usually far from cities, this also creates transport needs. Most of the increased health, food, water and energy demand is anticipated to occur in developing countries. The lack of availability of certain key resources such as fossil fuels, water and increasingly constrain the capacity to meet the health, food, energy and water demand in the Global South (UNEP, 2014).

The health-food-water-energy nexus as a framework for sustainable resource planning

The results from the study show that the health-energy-food-water is a conceptual framework presenting opportunities for greater resource coordination, management and policy convergence across sectors (Carter and Gulati, 2014). The absence of coordination among health, food, and energy and water systems results in trade-offs in planning which threaten the sustainability of development initiatives. The research findings also highlight the importance of the health-food-water-energy nexus in sustainably addressing issues of sectoral coordination of resources through harmonised institutions and policies (IRENA, 2015). Furthermore, the nexus also facilitates the setting of targets and indicators to direct and monitor developments in health, food, energy and water systems. An understanding of the nexus is also important in the promotion of inclusive development and transformation of vulnerable communities into resilient societies (FAO, 2014). Furthermore, the health-food-water nexus approach provides opportunities for coordinated planning which increase efficiencies in developing a nexus-sensitive agricultural plan. It also provides opportunities for assessing the synergies and trade-offs for reducing agricultural water use (Cilliers, 2008). However, the research findings also revealed that the application of the health-food-energy-water nexus approach is also limited due to the absence of models, indices, that can be useful in evaluating synergies and trade-offs (Sachs, 2010). The research findings revealed that the health-food-water-energy nexus is a multicentric approach which considers all its components as equal and interlinked. The nexus plays an important role in adapting to the challenges posed by population growth and climate variability and change, as well as in optimising resource use to achieve sustainable development goals (United Nations, 2013).

Discussion and Synthesis

The results revealed that the major drivers of the health-energy-food-water nexus are: economic growth, population expansion, urbanisation, geopolitics, technological development and climate change. This entails that there will be new pressures arising within the nexus which needs careful management. There is need for new responses to these challenges as well as changing health, food, energy and water systems such that they enhance resilience in countries of the Global South. Similarly, to note is that although there is a widespread recognition of the health, food, energy; water linkages management of these component systems is located in different sector specific institutions. This is partly as a result of policy environments in the Global South which fail to recognise cross sectoral linkages in resource management as policies and institutions are designed to work in silos thus creating an imbalance and duplication in the allocation of resources resulting in inefficiency in resource. Although the results from the study exposed the potential of health-food-energy-water nexus in facilitating regional integration between the countries in Southern Africa as espoused in the WEF Nexus Action Plan there is a lack of cross sectoral linkages between policies, institutions, policies and projects. This presents challenges in the implementation of the nexus approach as there are no guiding frameworks for its implementation. Furthermore, there is also a lack of clarity on the spatial scale under which implementation is supposed to occur.

Further, to note is that the pressure to produce more food and energy under increasing water scarcity requires the nexus to balance competing demands for water resources. In contrast to popular developmental approaches which have negatively impacted on other sectors and sustainability, the nexus approach could potentially contribute to sustainable socio-economic and inclusive development as it is multi-centric and holistic. The research findings also revealed that climate change could induce more pressure on health, food, water and energy systems which entail the need for having an integrated approach to sustainable development such as the health-food-water-energy nexus approach which promotes sustainable development thus ensuring socioeconomic security. Climate change projections which indicate increased pressure on health, water, energy and food resources further validate the need for a coordinated and integrated approach to sustainable development.

Conclusion and Policy Options

This study sought to explore the potential of recognising the synergies between health's, food, water and energy systems in building resilient cities in countries of the Global South. The research findings revealed that global population growth, urbanisation, climate change and changes in consumer patterns are drivers increasing demand for health, food, energy and water systems this does not only put pressure on these component systems but also increases the interconnections between these systems. The health-food-water-energy nexus approach has the potential to play an important role in adapting to the challenges posed by population growth and climate variability and change, as well as in optimising resource use to achieve sustainable development goals. However, it is also noted in this study that the policy environments in the Global South fail to recognise cross sectoral linkages in resource management as policies and institutions are designed to work in silos thus creating an imbalance and duplication in the allocation of resources resulting in inefficiency in resource. Furthermore, there is also absence of metrics and indicators for directing and monitoring developments in health, food, energy and water systems. An integrated model and tools for health-energy-water nexus implementation is recommended in this study as a means of promoting integrated regional development thus, ensuring socio-economic and political security and achieve regional integration. Furthermore, there is need for climate change policies to be taken from the perspective of the health-food-energy-water nexus as a way of avoiding maladaptation and negative externalities.

References

Carter, S., and Gulati, M. (2014). Understanding the Food Energy Water Nexus Climate change, the Food Energy Water Nexus and food security in South Africa. *WWF report.*

Chadwick, J. B. (2012). Citrus – Snapshot of the Present and a Look into the Future. Available online: http://www.cga.co.za/ pages/4710. [Accessed: 13 May 2015].

Chijioke, O. B., Haile, M., and Waschkeit, C. (2011). Implication of Climate Change on Crop Yield and Food Accessibility in Sub-Saharan Africa. Interdisciplinary Term Paper, ZEF Doctoral Studies Program, Bonn.

Cilliers, P. (2008). Complexity Theory as a General Framework for Sustainability Science in Burns, M. and Weaver, A. (Eds.). *Exploring Sustainability Science: A Southern African Perspective.* Stellenbosch: African Sun Media. pp. 39-57.

Cresswell, J, (2003). Research Design, Qualitative, Quantitative, Mixed Methods. Sage Publications

Fargione, J., Hill, J., Tilman, D., Polasky, S., and Hawthorne, P. (2008). Land Clearing and the Biofuel Carbon Debt. *Science* 319(5867):1235-1238.

FAO. (2006). Livestock's long shadow: environmental impacts and options. Available online: http://ftp.fao.org/docrep/ fao/010/a0701e/a0701e00.pdf. [Accessed: 13 May 2015].

FAO. (2008). Climate Change and Food Security: A Framework Document. Food and Agriculture Organisation of the United Nations (FAO), Rome.

FAO. (2014). The State of Food Insecurity in the World 2014. Rome: Food and Agriculture Organisation of the United Nations.

Foresight. 2011. The future of food and farming. Final Report. Available online: https://www.gov.uk/ government/uploads/system/uploads/attachment_data/ file/288329/11-546-future-of-food-and-farming-report. pdf. [Accessed: 5 March 2015].

Gitz, V., and Meybeck, A, (2012). Risks, vulnerabilities and resilience in a context of climate change. *Building resilience for adaptation to climate change in the agriculture sector,* 19 (23): 55-112.

HLPE. (2012). Food Security and Climate Change. A report by the High-Level Panel of Experts on Food Security and Nutrition of the Committee on World Food Security, Rome.

Hoff, H. (2011). Understanding the Nexus. Background paper for the Bonn 2011 Conference: The Water, Energy and Food Security Nexus. Stockholm Environment Institute, Stockholm.

IRENA. (2015). Renewable Energy in the Water, Energy and Food Nexus. Abu Dhabi: International Renewable Energy Association.

IPCC. (2013). Summary for Policymakers. In: Stocker, T. F., Qin, D., Plattner, M., Tignor, S.K., Boschung, J., Nauels, A., Xia, Y., Bex, V., and Midgley, P. M. (Eds.). Climate Change 2013: The Physical Science Basis. Contribution of Working Group I to the Fifth Assessment Report of the Intergovernmental Panel on Climate Change. Cambridge University Press, Cambridge, United Kingdom and New York, NY, USA. [IPCC-XXVI/Doc.4]

Norström, A. (2007). Planning for drinking water and sanitation in peri-urban areas. Swedish Water House Report, 21.

Organisation for Economic Co-operation and Development [OECD]. (2012). Environmental Outlook to 2050: Consequences of Inaction. Paris: OECD Publishing.

Porter, J. R., and Semenov, M. A. (2005). Crop responses to climatic variation. Philosophical Transactions of the Royal Society B: *Biological Sciences*, 360(1): pp. 2021-2035.

United Nations. (2013). World Population Prospects: The 2012 Revision. New York: United Nations.

UNEP. (2013). City-Level Decoupling: Urban Resource Flows and the Governance of Infrastructure Transitions. A Report of the Working Group on Cities of the International Resource Panel. Paris: United Nations Environment Programme.

Sachs, J. (2010). Monitoring the World's Agriculture Opinion. *Nature* 46(6):11-14.

UN-Water. (2014). United Nations World Water Development Report 2014. New York: UN-Water.

UNPD. (2008). World population prospects, the 2008 revision. New York: United Nations Population Division.

United Nations Population Division [UNPD]. (2007). World urbanisation prospects, the 2008 revision. New York: UNPD.

UN-Water. (2012). Status Report on the Application of Integrated Approaches to Water Resources Management. Available online: http//:www.un.org/waterforlifedecade/pdf/un_water_status_report_2012.pdf. [Accessed: 7 September 2015]

Rodriguez, D. J., Delgado, A., DeLaquil, P., and Sohns, A. (2013). Thirsty Energy. Water Paper no. 78923. Washington, D.C.: World Bank.

Tadesse, D. (2010). ISS Paper 220. Institute for Security Studies, Pretoria.

Wakeford, J. J., and De Wit, M. (2013). Oil Shock Mitigation Strategies for Developing Countries. Report commissioned by the United Kingdom Department for International Development. Stellenbosch, South Africa: *Sustainability Institute*. 12(7),121-252

Wakeford, J. J., Lagrange, S. M., and Kelly, C. (2016). Managing the Energy-Food-Water-Nexus in Developing Countries: Case Studies of Transition Governance. Stellenbosch University: Stellenbosch, South Africa.

WEF. (2013). *Global Risks 2013* (Eighth Edition). Geneva: World Economic Forum.

Wheeler, T. R., Craufurd, P. Q., Ellis, R. H., Porter, J. R., and Prasad, P. V. (2000). Temperature Variability and the Yield of Annual Crops. *Agriculture, Ecosystems & Environment*, 82(1-3), 159-167.

Smit, B., and Wandel, J. (2006). Adaptation, Adaptive Capacity and Vulnerability. *Global environmental change*, 16(3), 282-292.

Wlokas, H. (2008). The Impacts of Climate Change on Food Security and Health in Southern Africa. *Journal of Energy in Southern Africa* 19(1), 251-325.

Printed in the United States
By Bookmasters